Springer Theses

Recognizing Outstanding Ph.D. Research

Aims and Scope

The series "Springer Theses" brings together a selection of the very best Ph.D. theses from around the world and across the physical sciences. Nominated and endorsed by two recognized specialists, each published volume has been selected for its scientific excellence and the high impact of its contents for the pertinent field of research. For greater accessibility to non-specialists, the published versions include an extended introduction, as well as a foreword by the student's supervisor explaining the special relevance of the work for the field. As a whole, the series will provide a valuable resource both for newcomers to the research fields described, and for other scientists seeking detailed background information on special questions. Finally, it provides an accredited documentation of the valuable contributions made by today's younger generation of scientists.

Theses are accepted into the series by invited nomination only and must fulfill all of the following criteria

- They must be written in good English.
- The topic should fall within the confines of Chemistry, Physics, Earth Sciences, Engineering and related interdisciplinary fields such as Materials, Nanoscience, Chemical Engineering, Complex Systems and Biophysics.
- The work reported in the thesis must represent a significant scientific advance.
- If the thesis includes previously published material, permission to reproduce this must be gained from the respective copyright holder.
- They must have been examined and passed during the 12 months prior to nomination.
- Each thesis should include a foreword by the supervisor outlining the significance of its content.
- The theses should have a clearly defined structure including an introduction accessible to scientists not expert in that particular field.

More information about this series at http://www.springer.com/series/8790

Jing Li

Structural Optimization and Experimental Investigation of the Organic Rankine Cycle for Solar Thermal Power Generation

Doctoral Thesis accepted by
University of Science and Technology of China,
Hefei, China

Springer

Author
Dr. Jing Li
University of Science and
Technology of China
Hefei
China

Supervisor
Prof. Jie Ji
University of Science and
Technology of China
Hefei
China

Co-supervisor
Prof. Gang Pei
University of Science and
Technology of China
Hefei
China

ISSN 2190-5053 ISSN 2190-5061 (electronic)
Springer Theses
ISBN 978-3-662-51417-7 ISBN 978-3-662-45623-1 (eBook)
DOI 10.1007/978-3-662-45623-1

Springer Heidelberg New York Dordrecht London

Softcover reprint of the hardcover 1st edition 2015

Printed on acid-free paper

Springer-Verlag GmbH Berlin Heidelberg is part of Springer Science+Business Media
(www.springer.com)

Parts of this thesis have been published in the following articles:

Li J, Pei G, Ji J (2010) Optimization of low temperature solar thermal electric generation with Organic Rankine Cycle in different areas. Appl Energy 87:3355–3365 (Reproduced with Permission).

Li J, Pei G, Li Y, Ji J (2010) Novel design and simulation of a hybrid solar electricity system with Organic Rankine Cycle and PV cells. Int J Low Carbon Technol 5:223–230 (Reproduced with Permission).

Pei G, Li J, Ji J (2010) Working fluid selection for low temperature solar thermal power generation with two-stage collectors and heat storage units. In: Manyala R (ed) Solar collectors and panels, theory and applications. ISBN:978-953-307-142-8 (InTech) (Reproduced with Permission).

Pei G, Li J, Ji J (2010) Analysis of low temperature solar thermal electric generation using regenerative Organic Rankine Cycle. Appl Therm Eng 30:998–1004 (Reproduced with Permission).

Pei G, Li J, Li Y, Wang D, Ji J (2011) Construction and dynamic test of a small scale Organic Rankine Cycle. Energy 36:3215–3223 (Reproduced with Permission).

Pei G, Li J, Ji J (2011) Design and analysis of a novel low-temperature solar thermal electric system with two-stage collectors and heat storage units. Renewable Energy 36:2324–2333 (Reproduced with Permission).

Li J, Pei G, Li Y, Ji J (2011) Evaluation of external heat loss from a small-scale expander used in Organic Rankine Cycle. Appl Therm Eng 31:2694–2701 (Reproduced with Permission).

Li J, Pei G, Li Y, Wang D, Ji J (2012) Energetic and exergetic investigation of an organic Rankine cycle at different heat source temperatures. Energy 38:85–95 (Reproduced with Permission).

Li J, Pei G, Li Y, Wang D, Ji J (2013) Examination of the expander leaving loss in variable Organic Rankine Cycle operation. Energy Convers Manage 65:66–74 (Reproduced with Permission).

Li J, Pei G, Li Y, Ji J (2013) Analysis of a novel gravity driven Organic Rankine Cycle for small-scale cogeneration applications. Appl Energy 108:33–43 (Reproduced with Permission).

Li J, Pei G, Ji J, Bai X, Li P, Xia L (2014) Design of the ORC condensation temperature with respect to the expander characteristics for domestic CHP applications. Energy 77:579–590.

Supervisor's Foreword

The organic Rankine cycle (ORC) is one of the most promising technologies for low-/medium-grade heat to power conversion. Compared with the steam Rankine cycle, the ORC is capable to realize more efficient expander at low power and show better thermodynamic performance at low temperature, etc. The ORC has been successfully applied in waste heat recovery, biomass energy, and geothermal power generation. This technology in power range above 100 kW has reached a considerable degree of maturity.

Solar thermal power generation is a new application of the ORC. Solar ORC from tens of kWe to a few hundred kWe has great potential to meet the residential demand on heat and power. It has advantages over the highly concentrated solar power technology in regard to the easier energy collection and storage, and the ability to supply energy near the point of usage. It also has advantages over solar photovoltaic technology in regard to heat storage in replacement of battery storage, cost-effectiveness at relatively high power, and the potential cooperation with biomass resource.

However, one critical problem associated with solar ORC is the low power efficiency. In the past 7 years, Jing Li has been devoted to reducing the thermodynamic irreversibility and demonstrating the feasibility of the solar ORC. In this work, several innovative solutions are proposed, including solar ORC with collectors for direct vapor generation and solar ORC with PV module. The heat collection, storage, and power conversion are optimized. These systems are promising, and seem more reliable, flexible, efficient, and cost-effective than the conventional one. The design, construction, and test of a prototype are conducted by Dr. Li. Experimental investigation and thermodynamic analysis of the ORC under different cold reservoir temperatures are performed. Overall, this work contains valuable information on the solar ORC.

Hefei, July 2013

Prof. Jie Ji

Supervisor's Foreword

Acknowledgments

I offer the sincerest gratitude to my supervisor Prof. Jie Ji. He gave me the chance to study in his team of solar thermal conversion in April 2007. "Dive into solar energy investigation with vigor and gusto, otherwise you will just float around and get nowhere fast in the coming years," that is what he suggested at my early stage of research. He has supported me throughout my work on the ORC and CPC with his patience and knowledge while allowing me the room to work in my own way.

I also owe my gratitude to Prof. Gang Pei, my co-supervisor. It would not have been possible to complete my research on the ORC and CPC without his help. The ORC system in USTC involves painstaking work by him. The design and fabrication of the 3.5 kW turbo-expander, the high-speed gearbox, and the whole system involve many participants from different agencies and companies, such as Hangzhou Advance Gearbox Group Co., Ltd, XiangFan Hanghua Aero-Technology Applying Co., Ltd, Hefei General Machinery Research Institute, and 609 Institute of Aviation Industry Corporation of China. Prof. Pei has spent a lot of time for the coordination and technology improvement. Owing to his efforts, the ORC system has been successfully built and experiment investigation can thereby be carried out. One simply could not wish for a better co-supervisor.

Thanks are given to the National Natural Science Foundation of China (NSFC). The research on the ORC and CPC is supported by NSFC under Grant Nos. 51178442, 51206154, and 51378483.

Contents

Nomenclature

A	Area (m^2)
	Coefficient
$a - b$	Coefficients
B	Coefficient
c	Coefficient
C	Capacity (J/kg)
	Coefficient
	Concentration ratio
D	Diameter (m)
E	Exergy (J)
ex	Exergetic
G	Irradiation (W/m^2)
g	Gravity constant (m/s^2)
H	Enthalpy (J)
h	Specific enthalpy (J/kg)
	Heat transfer coefficient ($W/(m^2K)$)
k	Conductivity (W/(mK))
$L - P$	Coefficients
l	Length (m)
M	Molecular weight
m	Mass flow rate (kg/s)
N	Rotation speed (s^{-1})
Pr	Prandtl number
p	Pressure (Pa)
q	Heat (W)
R	Gas constant
r	Ratio
S	Area (m^2)
	Entropy (J/K)

T	Temperature (°C)
	Torque (Nm)
t	Time (s)
U	Heat transfer coefficient (W/(m^{2}K))
v	Kinematic viscosity (m^2/s)
	Specific volume (m^3/kg)
w	Power (W)
X	Mass fraction
	Dryness
x	Mole fraction
	Cover ratio

Greek Symbols

α	Absorptivity
	Coefficient
β	Volume coefficient of expansion (K^{-1})
δ	Volume ratio
ε	Machine efficiency
	Emissivity
γ	Activity coefficient
η	System efficiency
θ	Dimensionless
	Temperature
ρ	Density (kg/m^3)
	Reflectivity
π	Pressure ratio
σ	Stefan-Boltzmann constant
τ	Transmissivity
ζ	Exergy loss (J)
$\prod$	Osmotic pressure (Pa)

Scripts

0	Design point
a	Absorber
	Ambient
	Available exergy
air	Air
b	Binary phase
CHP	Combined heat and power
c	Collector
	Condenser

conv	Convective
cell	PV cell
e	Evaporator
eff	Effective
f	Working fluid
g	Generator
h	Hot side
i	Inlet
	Element
L	Glass cover
f	Working fluid
l	Liquid
	Low
net	Net power
ORC	Organic Rankine cycle
o	Outlet
p	Pump
	Pressure
rad	Radiative
R	Reflector
r	Regenerator
	Reference point
rot	Rotation
S	Absorber
SE	Shaft power
SPG	Solo power generation
s	Isentropic
sky	Sky
sys	System
su	Supply
t	Expander
total	Total
tube	Tube
u	Used exergy
v	Vapor
w	Water

Chapter 1
Gradual Progress in the Organic Rankine Cycle and Solar Thermal Power Generation

The organic Rankine cycle (ORC) is a technology for low-grade heat to power conversion. The ORC functions in a similar way as the conventional steam Rankine cycle. The principle is simple. The organic fluid is pumped into a heat exchanger where it's vaporized. The high pressure vapor flows through an expander and outputs technical work due to the pressure drop. The exhaust from the expander goes into the other heat exchanger where it's condensed, and then returns to the pump. Compared with the steam Rankine cycle, the ORC has several advantages, including:

(1) Easier realization of efficient turbomachinery at low power. A water molecule possesses polar covalent bonds between oxygen and hydrogen atoms. The latent heat of vaporization of water is much larger than commonly used organic fluids, as shown in Table 1.1. Given the cycle efficiency, the specific enthalpy drop of water through the expander shall be much larger. In the low power range from few kW_e to few hundreds of kW_e, the expander is expected to undergo very small mass flow rate when using the working fluid of water. There are many negative effects associated with the small flow rate, e.g. more significant friction losses due to smaller blade heights and passages, more appreciable leakage losses due to larger blade tip clearance relative to size, and secondary flow losses due to larger relative thickness of blades. These obstacles will lead to inefficient expansion, but can be overcome when using some appropriate organic fluids. Besides, the organic fluids can offer simpler configuration of the turbomachinery. On the condition of hot and cold side temperatures, the pressure ratio of the organic fluids is lower than that of water as shown in Table 1.2. Lower pressure ratio may avoid multistage expansion and complicated turbine.

(2) Excellent thermodynamic performance in utilization of low grade heat sources. Regulated by the slope of temperature-entropy (*T*-*s*) curve of the saturated vapor, the working fluids for the Rankine cycle can be divided into three categories: (a) dry fluids with positive slope, (b) wet fluids with negative slope, and (c) isentropic fluids with slope approximately equal to zero. Water is a typical wet fluid. To prevent water droplets from hitting and damaging the turbine blades at a high speed in the expansion process, superheat is generally

J. Li, *Structural Optimization and Experimental Investigation of the Organic Rankine Cycle for Solar Thermal Power Generation*, Springer Theses,
DOI 10.1007/978-3-662-45623-1_1

Table 1.1 Latent heat of water and organic fluids

Temperature (°C)	Latent heat (kJ/kg)				
	Water	R245fa	R123	Pentane	Benzene
80	2308.0	152.5	145.5	318.1	394.7
100	2256.4	134.5	134.0	296.4	379.7
120	2202.1	111.8	120.5	271.1	363.7
140	2144.3	78.8	104.0	240.8	346.6
160	2082.0	/	81.9	202.2	328.0
180	2014.2	/	40.6	147.0	307.4

Note / means supercritical state

Table 1.2 Ratio of the saturation pressure at certain temperature to that at 30 °C for water and organic fluids

Temperature (°C)	Pressure ratio				
	Water	R245fa	R123	Pentane	Benzene
80	11.2	4.4	4.5	4.5	6.4
100	23.9	7.1	7.2	7.2	11.3
120	46.8	10.8	10.9	11.1	18.9
140	85.1	15.9	16.0	16.2	29.7
160	145.6	/	22.7	23.0	44.7
180	236.1	/	31.5	31.8	64.6

Note The saturation pressure of water, R245fa, R123, pentane, benzene at 30 °C is 4.2, 177.8, 109.6, 82.0 and 15.9 kPa respectively

required prior to the expansion of the steam. The degree of superheat (ΔT) is illustrated in Fig. 1.1a and its values on different conditions are listed in Table 1.3. The peak temperature in the Rankine cycle is restricted by the heat source. Given the source conditions, a high degree of superheat reduces the average temperature of water during the heating process. The cycle efficiency is low according to Carnot's theorem. Through the replacement of water by dry fluids, this problem can be solved. Superheat is fended off as depicted in Fig. 1.1b. With a peak temperature of 150 °C and a condensation temperature of 35 °C, the efficiency of the ideal steam Rankine cycle with superheater is only 8.3 %, while it is 18.7 % when the working fluid is R245fa.

(3) Good reaction to low environment temperature. The pressure in the condenser is critical to the Rankine cycle efficiency as it is linked with the expander backpressure which limits the amount of energy drawn from the steam or organic fluids. A lower pressure in the condenser implies a larger specific enthalpy drop and a higher thermal efficiency. However, in normal operation it's very difficult to maintain a vacuum lower than 5 kPa [1]. As for water, the saturation temperature at 5 kPa is about 33 °C. When water is condensed below this temperature, no lower backpressure and no more power output can be

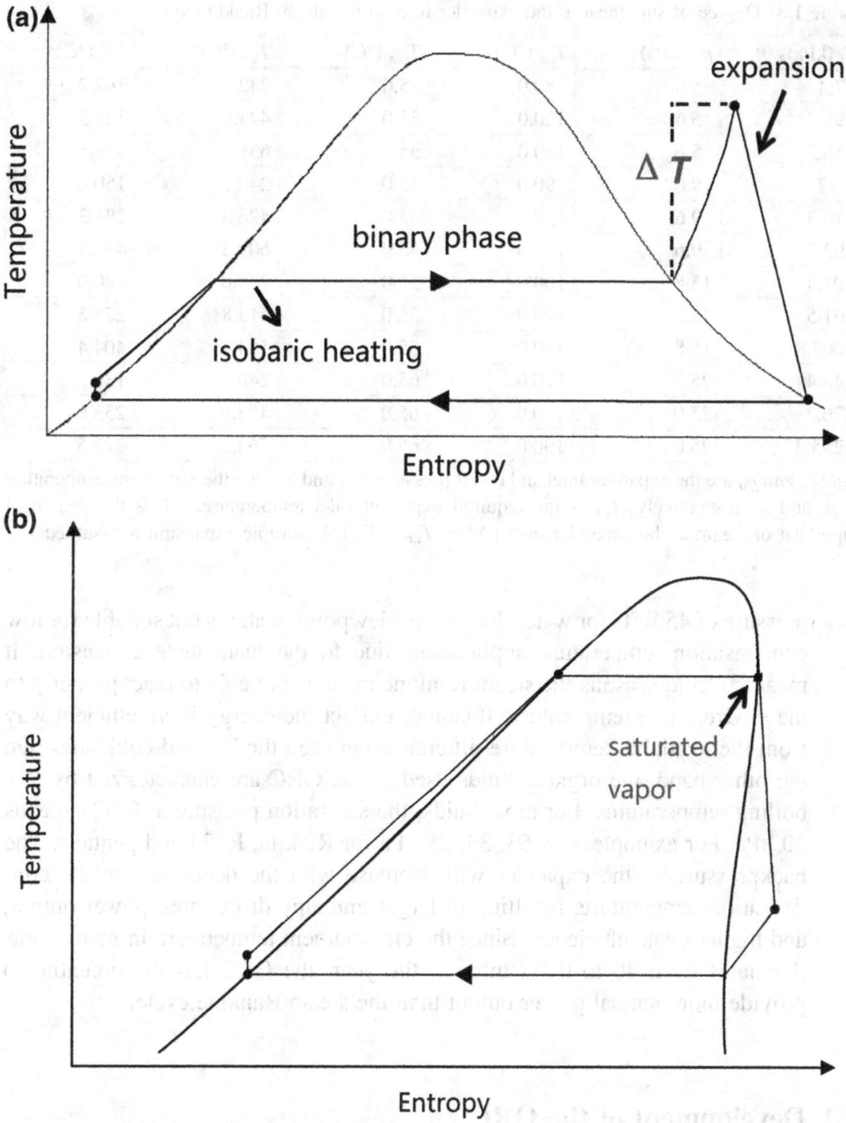

Fig. 1.1 The temperature—entropy curves: **a** water, **b** R245fa

realized for the expander. In this case, the power output does not increase with the decrement in the cold side temperature, which is in contrast to theoretical expectation. The cycle efficiency may be even decreased as more energy is needed to heat water at a lower temperature from the condenser. There are also pressure losses in the condenser, pipes, etc. The realizable backpressure of the expander in small power system shall be higher than 10 kPa, with a saturation

Table 1.3 Degree of superheat at the expander inlet in the steam Rankin cycle

p_i (kPa)	p_o (kPa)	$T_{i,s}$ (°C)	$T_{o,s}$ (°C)	$T_{i,r}$ (°C)	ΔT (°C)
47.4	5.6	80.0	35.0	242.2	162.2
198.7	5.6	120.0	35.0	441.5	321.5
618.2	5.6	160.0	35.0	638.5	478.5
70.2	9.6	90.0	45.0	240.6	150.6
270.3	9.6	130.0	45.0	426.3	296.3
792.2	9.6	170.0	45.0	609.2	439.2
101.4	15.8	100.0	55.0	240.0	140.0
361.5	15.8	140.0	55.0	413.8	273.8
1002.8	15.8	180.0	55.0	584.4	404.4
143.4	25.0	110.0	65.0	240.5	130.5
476.2	25.0	150.0	65.0	403.6	253.6
1255.2	25.0	190.0	65.0	563.5	373.5

Note p_i and p_o are the expander inlet and outlet pressure. $T_{i,s}$ and $T_{o,s}$ are the saturation temperature at p_i and p_o respectively. $T_{i,r}$ is the required expander inlet temperature. ΔT is the degree of superheat of steam at the expander inlet ($\Delta T = T_{i,r} - T_{i,s}$). Isentropic expansion is assumed

pressure of 45.8 °C for water. From this viewpoint, water is not suitable for low condensation temperature applications due to the inadequate expansion. It means in cold seasons the steam Rankine cycle is not able to react properly to the environment temperature. It cannot extract the exergy in an efficient way from the available temperature difference between the hot and cold sides. On the other hand, the organic fluids used in the ORC are characterized by low boiling temperatures. For most fluids, the saturation pressure at 0 °C exceeds 20 kPa. For example, it is 53, 33, 25 kPa for R245fa, R123 and pentane. The backpressure of the expander will decrease with the decrement in the condensation temperature, resulting in larger enthalpy drop, more power output, and higher cycle efficiency. Since the environment temperature in many areas fluctuates from 40 to 0 °C through the year, the ORC has the potential to provide more annual power output than the steam Rankine cycle.

1.1 Development of the ORC

The Rankine cycle was described in 1859 by William John Macquorn Rankine, a Scottish polymath and Glasgow University professor. This cycle, in the form of steam engines, has been very important for producing power since the 19th century, when it was used for transportation and industrial power generation. Nowadays it generates more than 85 % of all electric power used throughout the world. For the past decades, the Rankine cycle has evolved in a number of ways including the improvement of the steam turbine, refinement in configuration (reheat and

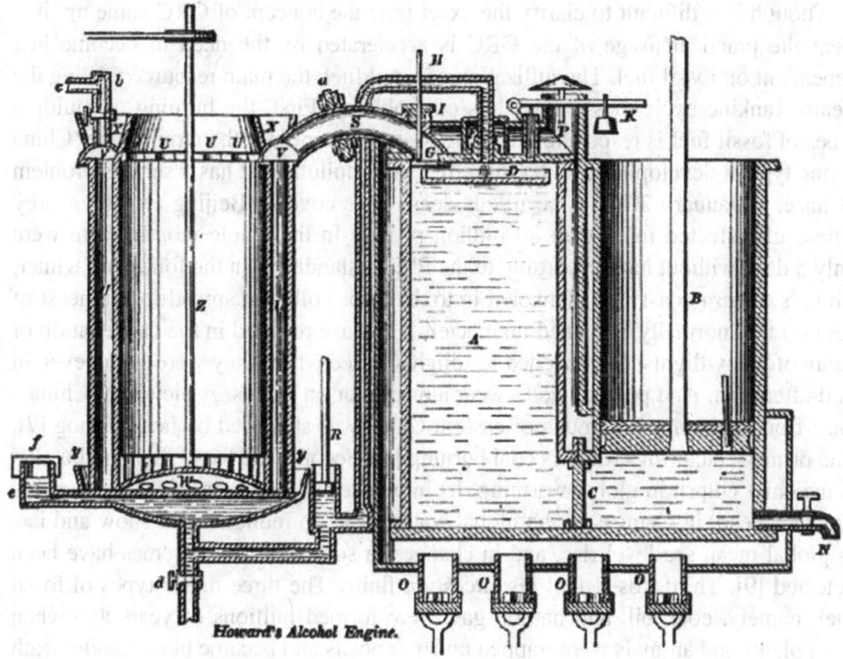

Fig. 1.2 Howard's alcohol engine [3]

regenerative features), and the use of different working fluids than water. Although the Rankine cycle using the organic fluids has never been as important as the steam Rankine cycle, the concept may be rather old. In the 18th and 19th centuries, water, alcohol, ammonia (some kind of 'organics'), together with mercury, air, etc. were commonly taken into consideration in the thermodynamic investigation on heat engines. A machine working with alcohol was proposed in 1797 by Cartwright. It was pointed out that due to the imperfect construction at the time, leakage of the vapor would make the system very uneconomic [2]. An engine which used the vapor of alcohol (Howard's alcohol engine) was patented in 1825 [3, 4]. A sketch of this engine is shown in Fig. 1.2. Attention was paid to minimize the leakage of the expensive working medium. An experiment with the mixture of water and methyl alcohol was performed in 1885, by running a launch engine 24 h with steam and 24 h with the mixture [5]. The mixture contained 15 % of wood alcohol. The results showed the cost of wood alcohol must be lower than 1/67 of its market price in order to enable the binary vapor to compete with water. An ammonia engine was devised by Lamm in 1869. The engine consisted of an inner tank, water vessel ammonia boiler, dome, delivery pipe, throttle, etc. It was tested by street railway companies in New Orleans with satisfactory results [6]. In short, many fluids were referred in the early heat engine inventions. The steam engine in the Rankine cycle was a widely used heat engine in the 19th century. It should be reasonable to consider different fluids when explicating the mechanisms of the Rankine cycle.

Though it is difficult to clarify the exact time the concept of ORC came up, it is clear the practical usage of the ORC is accelerated by the need to become less dependent on fossil fuel. The utilization of fossil fuel, the main resource feeding the steam Rankine cycle, has caused lots of problems. First, the burning of multiple types of fossil fuel is responsible for releasing pollutant into the atmosphere. China is one typical developing country suffering from pollutant. It has a serious problem of haze. In January 2013, a hazardous dense haze covered Beijing and the nearby cities, and affected more than 80 million people. In the whole month, there were only 5 days without haze according to the PM2.5 standards. In the following winter, China's eastern coast was blanketed in toxic haze. Pollution spread to southeast of country that normally remained unaffected. The haze resulted in the cancellation or delay of many flights. Cars needed headlights to see where they were going even in mid-afternoon. And people had to wear a respirator on streets. A picture of China's haze from space showed much of eastern China was shrouded by heavy smog [7]. The primary factor of PM2.5 is coal burning [8]. Second, the fossil fuel combustion is the chief culprit in global warming. Its influences on warming of the atmosphere and the ocean, in changes in the global water cycle, in reductions in snow and ice, in global mean sea level rise, and in changes in some climate extremes have been detected [9]. Third, fossil fuel resources are finite. The three major types of fossil fuel, namely, coal, oil, and natural gas, were formed millions of years ago when dead plants and animals were trapped under deposits and became buried underneath land. They cannot be replenished once used. As reported by BP, the proved reserves of coal, oil, natural gas were 860,938 million tons, 1668.9 billion barrels and 6614.1 trillion cubic feet at the end of 2012. The corresponding reserves/production (R/P) ratio was 109, 52.9 and 55.7 years [10].

Since the oil-shock in 1970s, there have been significant developments in the ORC. The advantage of ORC in low temperature and power applications is a good match for heat sources from the combustion of biomass, geothermal energy, and industrial exhaust, etc. Biomass is available worldwide. The collection, transportation, and storage costs motivate decentralized power generation. Though electricity generation from biomass using steam turbines is an extremely widespread bioenergy technology, the ORC systems are more suitable when the power capacity is below 3 MWe. The heat from the combustion of biomass is transferred from the exhaust gases to conduction oil through heat exchangers at a temperature ranging from 150 to 320 °C, and is supplied for the ORC. Exploitable geothermal resources are also abundant. Hydrothermal systems are traditionally classified as hot water fields (at temperatures up to 100 °C), wet steam fields (at temperatures exceeding 100 °C) and vapour-dominated fields (generally at temperature more than 200 °C). Less than 10 % of the hydrothermal systems are vapour-dominated. Most geothermal areas contain moderate-temperature water [11]. The potential for waste heat recovery from industrial processes is enormous. For example, about 45 % of America energy consumption is released to the atmosphere as waste heat. By recovering the waste heat 440 million tons/year of CO_2 emissions could be eliminated [12]. The steel making, cement kilns and glass making, etc. consume and create a large amount of energy. The exhausts in these sectors often leave a flue as hot as 300 °C and are discharged to the atmosphere. For a typical cement kiln line of 2,000 t/d capacity with a 4-stage cyclone preheater and grate cooler, assuming a preheater waste-gas temperature of 350 °C and grate cooler exhaust-air temperature of 275 °C, the annual loss to be attributed to unused process heat is approximately US $1.0–$1.6 million [13]. All these sources can drive the ORC and produce power.

Hundreds of ORC plants have operated reliably for many years all over the world. Some information is presented in Table 1.4. By the end of 2013, the installed capacity of ORC power plants has climbed to approximately 1,700 MW. And the growth gets faster and faster, as shown in Fig. 1.3. For most of the plants, the heat source temperature for the ORC is between 110 and 320 °C, with the net electric efficiency from 9 to 20 %. At present Turboden and Ormat are two the largest manufacturers in the field. The former has constructed more than 200 biomass ORCs and the latter has built up more than 30 geothermal ORCs. Triogen claims the most efficient ORC in the market with an efficiency of >17 % (150 kW) and also the most cost-effective one measured in cost per kW installed capacity. And the payback period is between 2–5 years [14]. The ORC technology in power range above 100 kW has reached a considerable degree of maturity.

1.2 Challenges for the ORC

The ORC power plants are gaining ever increasing interest. The market shall expand greatly in the coming years. However, there are challenges the ORC has to handle.

The industrial waste heat, geothermal energy, and biomass energy are the main sources for current commercial ORCs. The industrial processes are becoming more efficient. Many ways have been developed to increase the efficiency of industrial energy systems. Advanced boilers and furnaces can operate at higher temperatures while burning less fuel, which have efficiency more than 95 %—insulating flue stacks and installing heat exchangers that capture waste heat. The efficiencies of pumps and compressors, etc. depend on lots of factors but often improvements can be made by implementing better process control and better maintenance practices. Adjustment of the industry structure is also made from a strategic perspective to save energy. In industry of high energy consumption, the mainstream trend is the elimination of small and medium enterprises of low efficiency and high pollution. This trend can be reflected by the norm of energy consumption per unit products of cement of China, as shown in Table 1.5. It seems only large production lines can meet the ever-stricter standard. With enlarged capacity per unit and the increased efficiency in industrial processes, the quality of waste heat can be reduced. In such cases, the ORC may become uneconomic. Meanwhile, the ORC faces the competition from the steam Rankine cycle in recovering waste heat in large plants. The ORC has less technical superiority at higher power.

Table 1.4 Some ORC power plants in operation

Location	Started	Manufacturer	Working fluid	Heat source	Hot side temperature	Capacity
Wabuska, US	1984	Ormat	Pentane	Geothermal energy	104 °C	1.8 MW
Washoe, US	1986	Ormat	Pentane	Geothermal energy	/	78 MW (7 plants)
Empire, US	1987	Ormat	Pentane	Geothermal energy	118 °C	3.6 MW
Amatitlan, Guatemala	2007	Ormat	/	Geothermal energy	/	(12 + 12 + 1.2) MW
Imperial	1993	Ormat	Pentane	Geothermal energy		92 MW (2 plants)
Chena Hot Spring, USA	2007	UTC	R134a	Geothermal energy	74 °C	200 kW
Altheim, Austria	2001	Turboden	C_5F_{12}	Geothermal energy	106 °C	1 MW
Bad Mergentheim, Germany	2012	Turboden	Silicone oil	Biomass combustion	310 °C (in) 130 °C (out)	1 MW
Abbiategrasso (MI), Italy	2009	Turboden	Silicone oil	Biomass combustion	310 °C (in) 130 °C (out)	200 kW
Chivasso, Italy	2012	Turboden	Silicone oil	Waste heat	290 °C (in) 145 °C (out)	1 MW
London, UK	2011	Turboden	Silicone oil	Biomass combustion	310 °C (in) 130 °C (out)	1 MW
Gossau, Switzerland	2010	Turboden	Silicone oil	Biomass combustion	310 °C (in) 130 °C (out)	1 MW
Biessenhofen, Germany	2008	Turboden	Silicone oil	Biomass combustion	310 °C (in) 135 °C (out)	2 MW
Aleşd, Romania	2012	Turboden	Silicone oil	Waste heat	315 °C (in) 130 °C (out)	4 MW
Munksjö Aspa Bruk, Sweden	/	Opcon	R717	Waste heat	76–81 °C	Up to 580 kW
Stora Enso Skutskär Mill, Sweden	/	Opcon	R717	Waste heat	67–69 °C	Up to 500 kW

(continued)

Table 1.4 (continued)

Location	Started	Manufacturer	Working fluid	Heat source	Hot side temperature	Capacity
Scharnhauser Park, Germany	2003	GET	MDM	Biomass combustion	300 °C (in) 240 °C (out)	950 kW
Pontcharra France	2013	Exergy	Pentane	Waste heat	/	700 kW
ABS Acciaierie	2012	Exergy	/	Waste heat	/	1 MW
Pamukoren	2011	Exergy		Geothermal energy	/	22.5 MW
AdP, Portugal	2011	Triogen	Toluene	Waste heat	>300 °C	160 kW
Suez, France	2011	Triogen	Toluene	Waste heat	>300 °C	160 kW
Landfill, Germany	2011	Triogen	Toluene	Waste heat	>300 °C	160 kW
Kirchweidach, Germany	2012	Cryostar	R134a	Geothermal energy	/	7.5 MW
Vigliano Biellese, Italy	2011	GE CleanCycle™	R245fa	Biomass combustion	150 °C	125 kW
Brugnera, Italy	2012	GE CleanCycle™	R245fa	Biomass combustion	130 °C	125 kW
Manitoba, Canada	2012	GE CleanCycle™	R245fa	Biomass combustion	/	125 kW

Note The data are collected from the websites of the manufactures

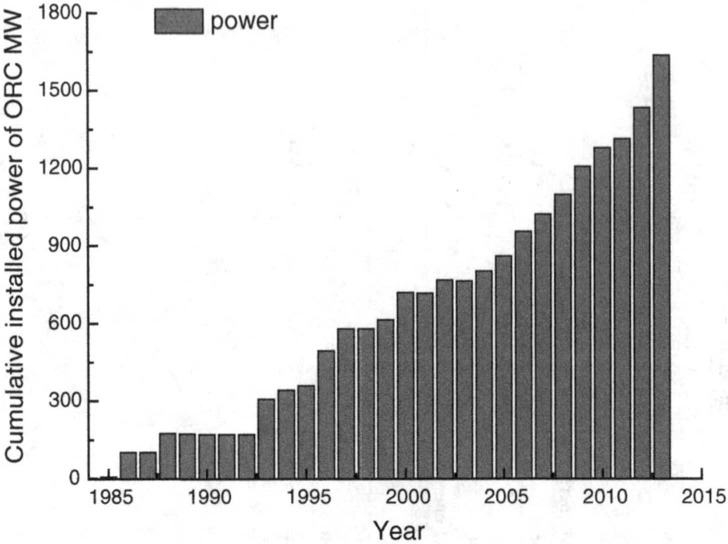

Fig. 1.3 Cumulative installed capacity of ORC plants. The data for recent years are obtained from websites of main manufactures including Ormat, Turboden, Exergy and Triogen

Table 1.5 Permissible range of energy consumption per unit products of cement (clinker) in China [16, 17]

	Plant	Year	Classification			
			Below 1,000 t/d	1,000–2,000 t/d	2,000–4,000 t/d	Above 4,000 t/d
Integrating standard coal consumption (kg/t)	E	2006	≤135	≤130	≤125	≤120
	E	2012	≤112	≤112	≤112	≤112
	N	2006	/	≤120	≤115	≤110
	N	2012	≤108	≤108	≤108	≤108
Integrating electricity consumption (kWh/t)	E	2006	≤78	≤76	≤73	≤68
	E	2012	≤64	≤64	≤64	≤64
	N	2006	/	≤68	≤65	≤62
	N	2012	≤60	≤60	≤60	≤60

Note E and N denote the existing and new built plants, respectively. 'Classification' denotes the capacity of production line. The unit t/d is the abbreviation of ton per day. / means the capacity is not allowed in the new built plant

Geothermal power is cost-effective, renewable, sustainable, and environmentally friendly, but has been limited to areas near tectonic plate boundaries. The temperature gradient drives a continuous conduction of thermal energy in the form of heat from the earth's core to the surface. Outside of the seasonal variations, the gradient through the crust is 20–40 °C per km of depth in most regions of the world, with low values at northerly latitudes. The potential capacity of power generation from geothermal energy increases with the increment in the depth of wells. To extend the market of geothermal ORCs, it is a must to extract geothermal fluids from the layers which underlie the reservoir now being exploited. As the wells get deeper the cost increases. Geothermal heat from thousands of meters deep could be promising, but the exploration and drilling may be prohibitively expensive.

The biomass is a very important renewable energy resource. And the market of biomass ORCs is growing extremely fast. The number of the total installed ORC units including biomass, geothermal, waste heat recovery plants in 2012 was about 370 [15]. While in February 2014, the number of under-construction biomass ORC units alone were at least 40 according to the data from the websites of the main manufactures. By cogeneration, the biomass ORC units can produce electricity and hot water for wood drying, sawdust drying, air pre-heating, district heating etc., thus meeting different types of consumer needs. However, many of current biomass ORCs are installed in companies that produce pellets, lumber, and furniture. It is easy to realize continuous and stable fuel supply in these applications. As the market extends, the supply chains will play a key role in cost-effective biomass ORCs. With the seasonality and low bulk density of most biomass feedstocks, the cost of transport and storage will be a big hurdle in the development of biomass ORCs.

1.3 ORC in the Solar Thermal Power Application

Compared with waste heat recovery, biomass and geothermal power generation, solar power generation is a new application of the ORC. Thermal power generation is one of the most important approaches in utilizing solar energy. More than 50 stations have been built during the past 5 years all over the world, including Solana Generating Station (USA, 280 MW, 2013), Solnova Solar Power Station (Spain, 150 MW, 2010), Welspun Solar MP project (India, 150 MW, 2014), Shams (United Arab Emirates, 100 MW, 2013), Hassi R'Mel integrated solar combined cycle power station (Algeria, 25 MW, 2011), Kuraymat Plant (Egypt, 20 MW, 2010), Archimede solar power plant (Italy, 5 MW, 2010), Jülich Solar Tower (Germany, 1.5 MW, 2008), Liddell Power Station Solar Steam Generator (Australia, 9 MW, 2012), Greenway CSP Mersin Solar Tower Plant (Turkey, 5 MW, 2013), Ain Beni Mathar Integrated Thermo Solar Combined Cycle Power Plant (Morocco, 20 MW, 2011), Thai Solar Energy (TSE) 1 (Thailand, 5 MW, 2011), Beijing Badaling Solar Tower (China, 1.5 MW, 2012) [18, 19]. And more than 20 station is under construction with overall capacity of about 2.5 GW [20]. Most of these operational and constructed stations using concentrating solar power (CSP) technologies such as parabolic trough, linear Fresnel reflector, heliostat tower, and dish/engine systems. The solar collectors are of high concentration ratio and the steam Rankine cycle is commonly adopted. Several disadvantages shall be noted, as follows:

(1) The technical difficulty in high temperature solar energy collection is great. Currently, over 95 % of the commercially operational solar thermal power plants are parabolic trough systems. A parabolic trough collector (PTC) is constructed as a long parabolic mirror with a linear receiver, running along its length at the focal point. The specially coated, evacuated receiver tube converts solar radiation into thermal energy, and its durability and efficiency is crucial for the sustainable profitability of the entire system. The glass-to-metal sealing of the receiver is a type of tubular sealing, which requires not only appropriate mechanical strength but also excellent gas tightness under high vacuum conditions. Due to the thoroughly different characteristics of metal and glass such as thermal expansion coefficient and wettability, sealing failure/degradation of the receiver may happen when the operating temperature fluctuates from about 400 °C at daytime to 30 °C at night. This failure/degradation has proven to be a big problem in the nine solar power plant (SEGS) in the USA [21]. Aside from the PTC, a parabolic dish collector is generally coupled with a small engine at the focus of the collector. By using many small engines in parallel it is easier to transport electrical energy than thermal energy from a field of dishes and reduce the energy lost in the transmission process. One example is the Maricopa Solar Plant in Arizona, USA. But when located at the focus of a dish, the engine makes the adjustment of the collector more difficult and casts shadows. For the heliostat, it has the disadvantage of relatively short life, and the mirror has to be manually realigned after any prolonged cloudy spell.

(2) A tracking system is required. In order to obtain the necessary temperature for the power conversion, solar radiation is concentrated in arrays. For most PTCs, the geometric concentration is more than 60 [22], and one-axis tracker is favorably used. The geometric concentration of parabolic dish collectors and heliostats can reach 1,000 or higher, and two-axis trackers are needed. Adding a solar tracking system means more moving parts and gears, which will call for regular maintenance and repair or replacement of broken parts. Also solar collectors with a tracker are prone to be damaged on extreme weather conditions compared with stationary collectors.

(3) Highly concentrated systems collect little diffuse radiation. Diffuse radiation describes the sunlight that has been scattered by molecules and particles in the atmosphere. Diffuse radiation does not have a definite direction. The part of diffuse radiation that can be utilized by a solar collector is inversely proportional to the concentration ratio [23]. The annual diffuse horizontal radiation and direct normal radiation have the same order of magnitude in lots of areas. For example, the ratio of the former and the latter in a typical meteorological year is 1:1.61, 1:0.95 and 1:0.86 for Beijing (N39°54′, E116°23′), Hefei (N31°51′, E117°16′) and Haikou(N20°01′, E110°19′) [24]. Given a concentration ratio higher than 60, the collector can hardly use any of the diffuse radiation. In this view, high-temperature solar thermal power generation is only applicable in certain regions of rich direct irradiation.

(4) A number of technical difficulties have to be overcome in high-temperature heat storage. Thermal storage is important to maintain the continuity of solar power generation. The large-scale utilization of solar energy is possible only if the effective technology for its storage can be developed with acceptable capital and running costs [25]. The first SEGS plant (SEGS I), built in 1984, included 3 h of heat storage. The plant used a mineral oil as the heat transfer fluid (HTF) and a two-tank thermal storage system: one held the cold oil and the other held the hot oil once it had been heated to about 300 °C. This technology proved to be successful for helping the plant dispatch its electric generation to meet the utility peak loads during non-sunlight periods. The mineral oil HTF was very flammable and could not be used for the later, more efficient SEGS plants that operated at higher solar field temperatures (approximately 400 °C). The HTF was also expensive, dramatically increasing the cost of larger HTF storage systems [26]. Latent heat storages are one possible HTF storage alternative, which are marked by a minimum of necessary storage material. Recently the world's largest solar thermal plant with molten salt storage system has come online in Arizona [27]. The 280 MW Solana Generating Station constructed by Spanish group Abengoa has 6 h of molten storage capacity. The storage system consists of six pairs of hot and cold tanks with a capacity of 125,000 metric tons of salt, and the molten salt is kept at a minimum temperature of 277 °C. However, the low heat conductivity of the salt is an obstacle which must be overcome to make full use of this storage technology.

(5) The plants have to be large to be economic. The capital cost per kilowatt of a solar power plant generally decreases with the increment in installed capacity [28]. The reason is the fixed costs for a smaller plant are approximately the same as that for a larger plant, and the cost per kW is higher. In past 5 years, many plants with capacity larger than 100 MW have been built. These plants typically occupy several square kilometers land or more for solar energy collection. Such large plants are likely to be restricted in remote regions and a consistent amount of sunlight must be available.

(6) There are potential negative impacts on environment along with large scale solar farm. Though solar thermal power plants produce carbon-free electricity, environmentalists have hit out at a giant new solar farm in the Mojave Desert. Mounting evidence has revealed birds flying through the extremely strong thermal flux surrounding the towers were scorched [29]. The mirrors may also cause vision problems for pilots. The reflection from the mirrors can be hazardous and the glare of light is so intense that it completely blinds the pilot as if looking into the sun [30].

By using the ORC, low-medium temperature solar thermal power system will be an attractive option that surmounts the above disadvantages. Since a temperature of about 100 °C or slightly higher is sufficient to drive the ORC, flat plate collectors (FPC), evacuated tube collectors (ETC) or small concentration ratio collectors will be competent in solar energy collection for the ORC. Particularly, compound parabolic concentrators (CPC) are capable to efficiently harness solar energy in temperature ranges from 100 to 150 °C. CPCs are non-imaging concentrators. In order to increase the efficiency of ETC collectors, a highly reflective, weather-proof CPC reflector is fitted behind, as shown in Fig. 1.4. The special, improved geometry of the CPC reflector ensures that direct and diffuse sunlight falls onto the absorber, which considerably increases the energy yield. CPC collectors with small concentration ratios can accept a large proportion of the diffuse radiation incident on their apertures, and direct it without tracking the sun [31]. At lower concentration ratios (e.g., 3X), CPC performance is substantially better than a double-glazed flat plate collector at temperature above 70 °C, while requiring only semi-annual adjustments for year-round operations [32]. The CPC collector exhibits excellent thermal performance for high-temperature applications such as steam generation, preparation and treatment of goods, in which operating temperature should be more than 120 °C [33, 34].

At present the CPC collectors have been mass-produced. Manufacturers have successfully developed CPC collectors with efficiency greater than 50 % on irradiation conditions of 800 W/m^2, working temperature of 130 °C, and environmental temperature of 30 °C [35]. With high-quality, corrosion-resistant and tested materials of the reflectors and tubes, the system can heat water to 150 °C [36]. New flat stationary evacuated CPC collectors with a target operating temperature adequate for industrial applications have been also developed, which exhibit proven system feasibility [37]. Different flat CPC collectors are available, reflecting the wide applicability of such technologies. And there are many industrial applications. The specifications of some commercial CPCs are listed in Table 1.6.

The low-medium temperature solar thermal electricity generation rooted to the ORC and CPC seems promising. The foreseeable merits include:

(1) CPC collectors with smaller concentration ratios can accept a large proportion of diffuse radiation incident on their apertures, and direct it without tracking the sun. Therefore, the cost associated with collectors can be reduced.

Fig. 1.4 CPC collectors

Table 1.6 Specifications of some CPC collectors [38, 39]

Ritter solar CPC	Series	CPC 12 OEM
	η_0 in relation to aperture (%)	64.2
	c_1 with wind, in relation to aperture (W/m² K)	0.89
	c_2 with wind, in relation to aperture (W/m² K²)	0.001
	Max. working overpressure, (bar)	10
	Max. stagnation temperature (°C)	272
	Collector material	Al/Cu/glass/silicone/PBT/EPDM/TE
	Glass tube material	Borosilicate glass 3.3
	Selective absorber coating material	Aluminum nitrite
Consolar CPC	Series	TUBO 12 CPC
	Test no. (ITW Stuttgart)	06COL457
	η_0, %	62
	c_1 (W/m² K)	0.395
	c_2 (W/(m² K²)	0.02
	Tube material	Borosilicate glass
	Transmission through glass (%)	92
	Vacuum	5×10^{-3} Pa (Getter: Barium)
	Lamination	SC layer (steel-copper-aluminum nitrite)
	Emission (%)	5–6
	Absorption (%)	93–94
	Max. perm. absorber temperature (°C)	250
	Reflector material	High-reflectivity aluminum

Note η_0 is the optical efficiency, c_1 and c_2 are the first and second heat loss coefficients

(2) The proposed solar ORC system enables scaling to smaller unit sizes. Electricity, as well as heating and cooling, can be supplied near its point of use, making it especially suitable for building-integrated solar energy applications.
(3) The technology of heat storage at temperature below 200 °C is much easier to realize compared with high-temperature heat storage. Many kinds of phase change materials can be used for the proposed system; these include paraffin, magnesium chloride hexahydrate, erythritol, galactitol, etc.
(4) The ORC is one of the most favorable and promising techniques in low temperature applications. ORC technology is reinforced by the high technological maturity of most of its components, stemming from their extensive use in refrigeration applications.

Another compelling motivation of the solar ORC is that it can cooperate with biomass energy. The physic characteristics of solar and biomass energy resources are similar: low energy density, seasonality and abundance. The hybrid solar/biomass ORC system can have three basic modes, as shown in Fig. 1.5. (I) Electricity is generated by solo solar ORC when irradiation is strong (V1 opens, V2 and V3 close). (II) Electricity is generated by solo biomass-fired ORC (V3 opens, V1 and V2 close). (III) The system is driven by solar energy as well as biomass energy when irradiation is available but not strong (V2 opens, V1 and V3 close).

The hybrid solar/biomass ORC system takes complementary advantages of the biomass-fired ORC and solar ORC. The biomass-fired system is a good thermal backup to keep the expander inlet temperature constant and to assist the solar collectors whenever irradiation is unavailable. On the other hand the collectors are useful for preheating the ORC fluid in the biomass-fired ORC process when irradiation is not strong enough for solo solar thermal electricity generation. In seasons of relatively high elevation angle of noon Sun, there is rich sunlight and photosynthesis is performed fast by plants in the hot/warm weather. The heat source of the ORC is mainly contributed by solar energy. Absorption chillers can also be driven by the heat source for cooling. In other seasons when rich radiation is unavailable, the chemical energy of the plants converted from solar energy can be released to fuel the ORC. With the coordination of the solar collector and biomass burner, a multifunctional ORC system of great flexibility in operation is realized. The capacity of storage units for solar and biomass energy can be considerably reduced or eliminated. The annual usage of the facility is increased and the payback period can be cut down. An illustration of this kind of system is presented in Fig. 1.6.

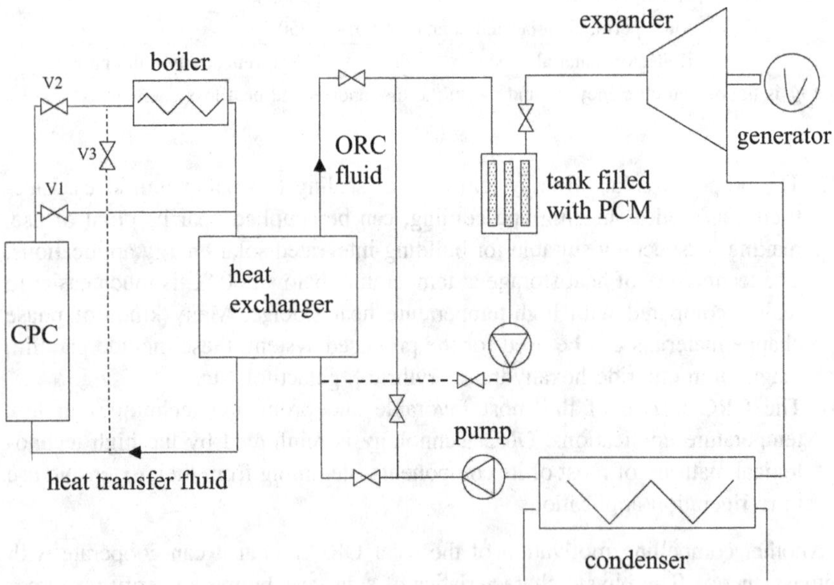

Fig. 1.5 Hybrid solar/biomass ORC system

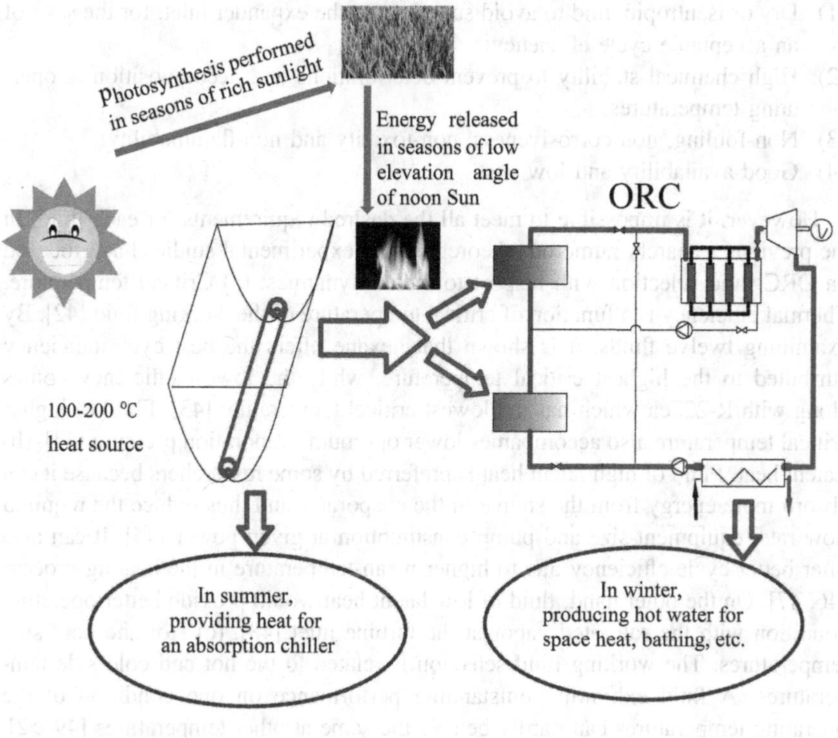

Fig. 1.6 A multifunctional ORC system

1.4 Recent Research in Low-Medium Temperature Solar Thermal Electricity Generation Using the ORC

The development of the low-medium temperature solar thermal power generation from 100 to 200 °C is subjected to the progress in ORC and non-tracking solar collector technologies. The following sections will focus on the recent research in these fields respectively.

1.4.1 ORC

In the past decade, ORC research has been very active in the working fluid selection, system design and optimization, low power expander design and test, etc.

The selection of the working fluid is the most important step for a successful ORC [40]. The reason is the fluid must have not only thermophysical properties that match the application but also adequate chemical stability at the desired operating temperature. There are some optimal characteristics of the working fluid [41]:

(1) Dry or isentropic fluid to avoid superheat at the expander inlet, for the sake of an acceptable cycle efficiency;
(2) High chemical stability to prevent deteriorations and decomposition at operating temperatures;
(3) Non-fouling, non-corrosiveness, non-toxicity and non-flammability;
(4) Good availability and low cost.

However, it is impossible to meet all the desired requirements for each fluid. In the previous research, numerous theoretical and experimental studies have focused on ORC fluid selection with respect to thermodynamics: (a) Critical temperature. Thermal efficiency is a function of critical temperature of the working fluid [42]. By examining twelve fluids, it is shown that hexane offers the best cycle efficiency attributed to the highest critical temperature, while the lowest efficiency comes along with R-227ea which has the lowest critical temperature [43]. Fluid of higher critical temperature also accompanies lower optimum evaporation pressure [44]. (b) Latent heat. Fluid of high latent heat is preferred by some researchers because it can absorb more energy from the source in the evaporator and thus reduce the required flow rate, equipment size and pump consumption at given power [45]. It can also offer better cycle efficiency due to higher mean-temperature in the heating process [46, 47]. On the other hand, fluid of low latent heat would provide better operating condition with the saturated vapor at the turbine inlet [48]. (c) Hot and cold side temperatures. The working fluid selection is related to the hot and cold side temperatures. A fluid exhibiting outstanding performance on one condition of the operating temperatures can hardly behave the same at other temperatures [49–52]. (d) Configuration of the thermodynamic cycle [53–56]. The working fluid selection for the ORC with an internal heat exchanger or intermediate heat exchanger or preheater is different from that for a basic ORC. (e) Other indicators as power output, cost, mass flow rate, efficiency and pinch point temperature [57–60]. There are generally several indicators rather than one to measure the ORC. A multi-objective function in comprehensive consideration of the properties and performances is more favorite in the fluid selection. Notably, there may be one hundred or more candidate fluids for the ORC, but in practical systems the fluids are limited. Siloxanes, toluene, pentane, and their derivatives are favorably used as working fluids in high-temperature applications. While in low-medium temperature applications, R245fa, R123 and ammonia are commonly utilized.

Thermodynamic optimization is also widely conducted, including: (a) Parametric optimization. Via optimizing the degree of sub-cooling of organic fluid at the condenser outlet and the mass flow rate of cooling water, the net power is increased [61]. Choosing a proper nominal state can improve the system output power and efficiency [62]. Given the turbine inlet pressure, the system irreversibility increases with the increment in the inlet turbine temperature [63]. (b) Structure optimization. Employing regenerator [64, 65], ejector [66] and two-stage expanders [67] results in higher cycle efficiency.

The expander is the key component of small-scale ORC systems. Yamanoto et al. designed and tested an experimental ORC with a micro-turbine and nozzle to determine the optimum design of a turbine blade shape. The maximum rotation speed and turbine output was 35,000 rpm and 150 W, respectively. The authors concluded that the ORC could be applied to low-grade heat sources and R123 was able to improve ORC performance significantly [48]. Badr et al. created a prototype expander by modifying an existing multi-vane pump. The expander worked through a reverse pumping process, thereby achieving an isentropic efficiency of over 73 % [68]. James et al. carried out experimental investigation on relatively cost-effective gerotor and scroll expanders, which generated 2.07 and 2.96 kW, and had isentropic efficiencies of 0.85 and 0.83, respectively. Both expanders demonstrated significant potential to produce power from low-grade energy [69]. Lemort et al. performed an experimental study on the prototype of an open-drive, oil-free scroll expander integrated into an ORC that ran on refrigerant R123. The maximum delivered shaft power was 1.82 kW, and the maximum achieved overall isentropic effectiveness was 68 %. Internal leakages and, to a lesser extent, supply pressure drops and mechanical losses, were the main losses affecting the performance of the expander [70, 71].

Scroll, screw and turbo expanders are the three main types of expanders used in the ORC system. Scroll expanders seem to be good candidate for the expansion device of small scale ORC systems, attributed to the simplicity of operation, reliability (no valves and few moving parts), low rotational speed and ability to handle high pressure ratios. At present most of the employed scroll expanders are obtained by modifying existing compressors. And some experimental works on this kind of expander are listed in Table 1.7. Screw expanders especially twin screw expanders show higher degree of technical maturity compared with scroll expanders, and can be commercially produced by companies such as Opcon Group and Jiangxi Huadian Electrical Power. In general, this kind of expander consists of a pair of helical screws and the shell casing. The fluid, which can be at superheated state, saturated state, binary state or liquid state, moves through several grooves and drives the twin helical screws in opposite directions. The expander has high tolerance for the fluid fluctuation in a wide range of pressure and temperature and volumetric flow. It also has other advantages as quick start-up and shut-down operation, no special warm-up, less faults from over-speeding and turning. Turbo-expanders have the highest degree of technical maturity and offer many advantages such as compact structure, light weight, small size, good stability and superior efficiency. Owing to the technical maturity, the turbo-expanders

Table 1.7 Some experimental works based on scroll expanders

Participant	Working fluid	Expander efficiency	Cycle electric efficiency (%)
Peterson et al.	R123	0.40–0.50	7.2
Kane et al.	R123/R134a	0.50–0.67	14.1
Manolakos et al.	R134a	0.30–0.50	3.5–5.0
Wang et al.	R245fa	0.45	5.8
Lemort et al.	R123	0.42–0.68	Unavailable

References [70, 75–77, 111]

are preferred in the existing ORC plants, for which the capacity is generally larger than 0.5 MWe. In low power application, the advantages of the turbo-expanders over the scroll and screw expanders become less appreciable regarding the requirement of multistage expansion or a gearbox to reduce the rotation speed. Scroll and screw expanders are therefore competitive at low power.

In the recent ORC studies, special attention was paid to small-scale power generation. Liu et al. developed and evaluated a biomass-fired micro-scale CHP system. The system generated 284 W electricity, which corresponds to 1.34 % electrical efficiency and 88 % overall CHP efficiency [72]. Riffat and Zhao designed and tested a 1.34 kW ORC-based CHP system assisted by fuel gas. The electrical efficiency was 16 %, and the overall efficiency was about 59 %. Further analysis showed that the proposed system would save primary energy of approximately 3,150 kWh per annum compared with conventional electricity and heating supply systems. And the energy savings would, in turn, result in reduced CO_2 emissions of up to 600 tons per annum [73, 74]. Kane et al. provided a novel concept of a mini-hybrid solar power plant that integrates a field of solar concentrators. Laboratory tests were conducted with two superposed ORCs. The chosen fluids were R123 for the topping cycle, and R134a for the bottoming cycle. An overall superposed cycle efficiency of 14.1 % was achieved for a supply temperature of up to 165 °C [75]. Peterson et al. presented a study on the performance of a small-scale regenerative Rankine power cycle that employed a scroll expander. The system efficiency was 7.2 % [76]. Manolakos et al. presented the detailed laboratory experimental results of a low-temperature ORC engine coupled with a reverse osmosis desalination unit. The results indicated that the efficiency of the Rankine cycle fluctuated from 3.5 to 5.0 % [77]. Gang et al. examined an innovative ORC system with two-stage evaporators, as well as a regenerative cycle suitable for domestic applications. System performance was estimated based on the commercial expander. The thermodynamic irreversibility could be reduced by using two-stage evaporators, and ORC efficiency could be increased by the regenerative cycle [78, 79]. Al-Sulaiman et al. carried out the energy and exergy analysis of a biomass tri-generation system using the ORC. A heating process and a single-effect absorption chiller were connected with the ORC condenser. It revealed that there was a significant improvement when tri-generation was used, for which the fuel utilization efficiency went up to 88 %, as compared with 12 % for solo electrical power [80]. Wang et al. introduced a novel thermally activated cooling concept by coupling an ORC and a vapor compression cycle (VCC). With both subcooling and cooling recuperation in the vapor compression cycle, the overall cycle COP reached 0.66 at extreme military conditions with outdoor temperature of 48.9 °C [81]. Bu et al. combined the ORC with VCC for ice making driven by solar energy. In terms of overall efficiency and ice production per square meter collector per day, R123 seemed the most suitable working fluid for the ORC/VCC [82]. Jradi and Riffat conducted the experiment investigation on an innovative micro-scale tri-generation system consisting of an ORC-based combined heat and power unit and a combined dehumidification and cooling unit. It was shown that the proposed system was capable of providing about 9.6 kW heating power, 6.5 kW cooling power and 500 W electric power [83].

Above all, small scale ORC and multifunctional systems are the new research trend. The interest in these systems is strengthened by the following aspects: (1) The size of the ORC plant is limited by the low energy density of heat sources. Biomass typically contains more than 70 % air and void space, and is difficult to collect, ship, and store. Solar radiation is generally less than 1,000 W/m^2, and a large area for gathering an appreciable amount of energy is not easily accessible. Yet, more than 90 % of available waste heat worldwide is applicable to the 10–250 kW system size [84]. (2) The size of the ORC plant is also limited by the availability of energy consumers. Many residential applications require only several to tens of kW for pumping, refrigerator, air conditioning, etc. (3) The small-scale production of electricity at or near customers' homes and businesses can improve the reliability of power supply. (4) Smaller modular systems may be more easily distributed in the market because of the possibility of using waste heat as well as electricity, and offsetting retail electricity costs instead of fossil plant generating costs [85]. In view of the cost-effectiveness, multifunctional systems are crucial to increase the output per unit of resource consumption, and to shorten the payback time.

1.4.2 Solar Collectors

CPCs are highly attractive because they do not require complicated tracking system and are able to collect solar radiation at a wider angular region of the sky, which includes a substantial portion of diffuse radiation. Research on CPC collectors for 100–200 °C applications is expanding, where the solar heat can be utilized directly for solar thermal power generation as well as industrial processes, desalination and solar cooling. There have been significant improvement in material selection and structure design for the CPCs. Frank et al. developed a non-tracking, flat, low-concentrating CPC collector for the economical supply of solar process heat at temperatures between 120 and 150 °C, in which the basic concept is the integration of an absorber tube and reflectors inside a low-pressure enclosure. A prototype, with an aperture area of 2.0 m^2, was tested and showed efficiencies of about 50 % on the conditions of operating temperature of 150 °C, radiation of 1,000 W/m^2 (900 W/m^2 direct) and ambient temperature of 20 °C [86]. Gudekar et al. presented a working model of CPC collector for steam generation at temperature up to 150 °C. An experimental demonstration unit having an aperture area of nearly 30 m^2, acceptance angle of 6°, requiring tilt adjustments once a day for a daily operation of 6 h was set up and tested. The results showed the CPC system had the potential of improving thermal efficiency up to 71 % [33, 87]. Sagade et al. investigated the effect of receiver temperature on performance evaluation of silver coated selective surface CPC with top glass cover. The line focusing parabolic trough yielded instantaneous efficiency of 60 % with top cover [88]. Liu et al.

introduced a novel all-glass evacuated tubular solar steam generator with simplified CPC. Outdoor experiments were carried out to study the actual performance of the generator in summer under different operating conditions. The results showed that the maximum steam outlet temperature exceeded 200 °C with pressure of 0.55 MPa. The solar steam generator can steadily produce steam over 150 °C in the pressure range from 0.1 to 0.55 MPa with a collecting efficiency over 0.30 [89]. Fernández and Dieste studied a low and medium temperature solar thermal collector based on innovative materials and improved heat exchange performance. The traditional metallic materials were replaced by surface treated Aluminum with TiNOx for the absorber, which simplified the production and assembly process. The definitive prototype had an aperture area of 0.225 m^2, and the accumulated efficiency was between 41 and 57 % [90]. Pei et al. tested a CPC-type solar water heater system with a U-pipe. Through the experimental study and exergetic analysis of the solar water heater system, thermal efficiency of above 49.0 % (attaining 95 °C water temperature) and exergetic efficiency of above 4.62 % (attaining 55 °C water temperature) can be achieved [91]. Li et al. developed and tested two truncated CPC solar collectors, which combined the external CPC and the U-shape evacuated tube. The two collectors had concentration ratios of 3.06 and 6.03 respectively. Experimental results indicated that the tilt angle of the 3X CPC collector did not need daily adjustment, while the 6X CPC collector needed to be adjusted five times a day. And the daily thermal efficiencies of the 3X and the 6X CPC collectors were 40 and 46 % respectively at the collecting temperature of 200 °C [92].

Aside from the purpose of medium temperature solar thermal applications, the need of reducing cost of photovoltaic (PV) cells has been driving the development of CPCs [93, 94]. By concentrating solar radiation onto a smaller area, the sizes of PV modules can be largely reduced, resulting in a remarkable cost reduction of the modules. For example, Mallick et al. presented a detailed comparative experimental characterization of an asymmetric CPC which was suitable for vertical building facade integration in the UK [94]. Development of the low concentration solar concentrators had been also based on refraction of lens [95], in which the functions and advantages of both the lens and CPC collectors were combined [96]. The lens-walled CPC had a thin lens-shape wall with mirror coating on its back surface and larger half acceptance angle for a given concentration ratio, which could result in the reduction in annual adjustments of the CPC tilt angle, making the device suitable for building-integrated applications [97]. The analysis with software PHOTOPIA showed that a lens-walled CPC with a geometrical concentration ratio of 4 had advantages over a common CPC with a geometrical concentration ratio of 2.5 in terms of actual acceptance angle, optical efficiency and optical concentration ratio [98]. The experimental results showed that the fill-factor of the mirror CPC dropped more sharply than that of the lens-walled CPC, which indicated that the latter had a more uniform flux distribution on PV [99]. The design methodology of the lens-walled CPCs may be useful for enlarging the acceptance angle in the medium temperature solar heat collections.

1.4.3 Coupling the ORC and Solar Collectors

The solar ORC system has been investigated intensively by researchers in the field of solar reverse osmosis (RO), with respect to parametric optimization [100], performance assessment [58, 101, 102], economic assessment [103, 104] and configuration design [105, 106]. For example, Delgado-Torres and García-Rodríguez gave the design recommendations for solar RO desalination based on ORC. The selection of the working fluid and boundary conditions of the ORC, operation temperature and configuration of the solar field, solar collector and thermal energy storage technology were discussed. It was recommended that Siloxane MM could be a good choice for ORC driven by PTCs and other fluids as Solkatherm® SES36, isobutene, isopentane, R245ca and R245fa for ORC driven by stationary solar collectors [107]. Manolakos et al. carried out on site experimental evaluation of a low-temperature solar ORC system for RO desalination. The system consisted of 54 vacuum tube collectors (Thermomax model TDS 300) of a total gross area of 216 m^2, and R134a was used as the working fluid. The results showed the average efficiency of the Rankine engine was 0.73 and 1.17 % for the cloudy and sunny day. Despite of the low efficiency, the system operated reliably during the experimental tests [108]. Nafey and Sharaf performed the design and performance calculations of the solar ORC using MatLab/SimuLink computational environment. The system consisted of FPC/PTC/CPC solar collectors for heat input, turbine for work output, condenser unit for heat rejection, pump unit, and RO unit. Different fluids were investigated. Toluene and water seemed favorable from the points of total solar collector area, specific total cost and the rate of exergy destruction [109].

There are also studies in solar ORC for other purposes than desalination. Quoilin et al. conducted the performance and design optimization of a low-cost solar ORC for remote power generation with PTCs. With conservative hypotheses and real expander efficiency curves, an overall electrical efficiency between 7 and 8 % could be reached [110]. Wang et al. designed, constructed, and tested a prototype low-temperature solar ORC system. The ORC and solar collectors worked independently. The ETC and PFC collectors were used, with tested efficiency of 71.6 and 55.2 % respectively. A rolling-piston expander was adopted and the average expansion work of the expander was 1.73 kW on using R245fa. The overall power generation efficiency was estimated at 4.2 and 3.2 % for ETC and PFC collectors [111]. He et al. built a model for a typical parabolic trough solar thermal power generation system with ORC within the transient energy simulation package TRNSYS. The influences of the interlayer pressure between the absorber and glass tubes, the flow rate of high temperature oil inside the absorber tube, solar radiation intensity and incidence angle, on the performance of the PTC field were analyzed. The heat loss of the solar collector increased sharply with the increase in the interlayer pressure at beginning and then climbed to an approximately constant value [112]. Wang et al. carried out the off-design performance analysis of a solar-powered ORC. The results indicated that a decrease in environment temperature, or the increases in thermal oil mass flow rates of vapor generator could improve the performance [113].

On the whole, research on the multifunctional ORCs and medium temperature collectors grows rapidly. More progress can be expected in the future. As a promising renewable technology, the solar ORC is attracting increasing interest and the room for improvement is large [114–117].

1.5 Scientific Issues and Innovative Features of the Thesis

The thesis investigates and develops the solar heat and power generation system that combines the advantages of the ORC and CPC. The ORC is driven by solar energy in the temperature range from 100 to 200 °C. The main scientific issues that the thesis aims to address are as follows:

(1) Optimization of the thermodynamic cycle. The application and performance of the ORC have been investigated by previous researchers; however, most of the investigations were focused on the ORC with one-stage evaporator, wherein a large sum of exergy was lost in the heat exchanger due to the unmatched temperature between the organic fluid and the conduction oil. And the pump in small scale ORC was characterized by low efficiency and there was considerable thermodynamic irreversibility in the pressurization process. To solve these problems in the conventional solar ORC system, innovative configurations are proposed.
(2) The identification of the performance advantages and disadvantages of small-scale expanders on using organic working fluids rather than water. Although small-scale ORC units in power ranges below 100 kW exhibit significant potential for electricity generation and heat supply at or near the site of consumption, the feasibility of small-scale expanders has yet to be demonstrated. Thorough and comprehensive studies on the performance of small-scale expanders especially turbo-expanders are essential.
(3) Experimental investigation and thermodynamic analysis of the ORC under different cold reservoir temperatures. Thermodynamics of the ORC on different cold reservoir conditions is the key scientific issue that is fundamental to the CHP performance evaluation and optimization. The ORC cold reservoir temperature varies with environment temperature and the consumer's demand on heat and power. It influences the ORC power conversion efficiency, available energy of the output heat, expansion ratio of the organic fluid at the expander inlet and outlet, net positive suction head in the pumping process and heat transfer temperature difference in the exchangers. It is difficult to predict the cycle performance on the variable operating conditions. At present, research on this topic is rare.

References

1. Bejan A (2006) Advanced engineering thermodynamics, 3rd edn. Wiley, Hoboken
2. The French Journal Nature, 21 June, 1890
3. Knight's practical dictionary of mechanics. Cassell & Co, London, Datum (1884)
4. Alcohol Engines. www.douglas-self.com/MUSEUM/POWER/alcohol/alcohol.htm. Accessed 16 April 2014
5. British Journal Engineering, 9 January, 1885
6. Appleton's Cyclopedia of American Biography, New York, 1887
7. China's dangerous smog seen from space. NOAA Environmental Visualization Laboratory. www.nnvl.noaa.gov/MediaDetail2.php?MediaID=1443&MediaTypeID=1. Accessed 22 Oct 2013
8. China Meteorological Administration. www.cma.gov.cn/2011zwxx/2011zyjgl/2011zyjgldt/201402/t20140224_239174.html. Accessed 11 March 2014
9. Intergovernmental Panel on Climate Change (IPCC), Climate Change (2013) The physical science basis—summary for policymakers, observed changes in the climate system, pp 10 and 11. In: IPCC AR5 WG1 2013
10. BP (British Petroleum) Statistical Review of World Energy. www.bp.com. Accessed 5 Dec 2013
11. Barbier E (2002) Geothermal energy technology and current status: an overview. Renew Sustain Energy Rev 6:3–65
12. Facts about waste heat. www.industrialwasteheat.com. Accessed 10 June 2014
13. Bronicki LY (2014) Organic Rankine cycle power plant for waste heat recovery. www.ormat.com/research/papers/organic-rankine-cycle-power-plant-waste-heat-recovery. Accessed 10 May 2014
14. Benefits of Triogen. www.triogen.nl/why-triogen/benefits. Accessed 12 May 2014
15. Colonna P, Larjola J, Uusitalo A, Saaresti T, Honkatukia J, Casati E, Mathijssen T, Trapp C (2013) Organic rankine cycle power systems: the path from the concept to current applications and an outlook to the future. In: 2nd international seminar on ORC power systems, Rotterdam
16. The norm of energy consumption per unit products of cement. National Standards of the People's Republic of China. GB16780-200X
17. The norm of energy consumption per unit products of cement. National Standards of the People's Republic of China. GB 16780-2012
18. Concentrating solar power projects. www.nrel.gov/csp/solarpaces/. Accessed 22 March 2014
19. List of solar thermal power stations. www.en.wikipedia.org/wiki/List_of_solar_thermal_power_stations#cite_note-63. Accessed 28 March 2014
20. Concentrating solar power projects under construction. www.nrel.gov/csp/solarpaces/projects_by_status.cfm?status=Under%20Construction. Accessed 25 March 2014
21. Dongqiang L, Zhifeng W, Fengli D (2007) The glass-to-metal sealing process in parabolic trough solar receivers. In: Proceedings of ISES world congress 2007, pp 740–744
22. Price H, Lupfert E, Kearney D, Zarza E, Cohen G, Gee R (2002) Advances in parabolic trough solar power technology. J Sol Energy Eng 124:109–125
23. Ajona JI, Vidal A (2000) The use of CPC collectors for detoxification of contaminated water: design, construction and preliminary results. Sol Energy 68:109–120
24. Energyplus. http://apps1.eere.energy.gov/buildings/energyplus/energyplus_about.cfm. Accessed 15 June 2014
25. Kenisarin M, Mahkamov K (2007) Solar energy storage using phase change materials. Renew Sustain Energy Rev 11:1913–1965
26. Herrmann U, Kelly B, Price H (2004) Two-tank molten salt storage for parabolic trough solar power plants. Energy 29:883–893
27. World's largest solar thermal plant with storage comes online. http://cleantechnica.com/2013/10/14/worlds-largest-solar-thermal-plant-torage-comes-online/. Accessed 3 Jan 2014

28. Prabhu E (2006) Solar trough organic Rankine electricity system (STORES) stage 1: power plant optimization and economics. Subcontract Report NREL/SR-550-39433, March 2006
29. Horror at the world's largest solar farm days after it opens as it is revealed panels are scorching birds that fly over them, mail online. www.dailymail.co.uk/news/article-2560494/Worlds-largest-solar-farm-SCORCHING-BIRDS-fly-it.html. Accessed 18 March 2014
30. Despite the complains made, pilots are still blinded by glare from solar power plant. http://mostepicstuff.com/despite-the-complains-made-pilots-are-still-blinded-by-glare-from-solar-power-plant/. Accessed 20 March 2014
31. Pereira M (1985) Design and performance of a novel non-evacuated1.2x CPC type concentrator. In: Proceedings of intersol biennial, congress of ISES, Montreal, Canada, 1985, vol 2, pp 1199–1204
32. Rabl A, O'Gallagher J, Winston R (1980) Design and test of non-evacuated solar collectors with compound parabolic concentrators. Sol Energy 25:335–351
33. Gudekar AS, Jadhav AS, Panse SV, Joshi JB, Pandit AB (2013) Cost effective design of compound parabolic collector for steam generation. Sol Energy 90:43–50
34. Saitoh TS (2002) Proposed solar Rankine cycle system with phase change steam accumulator and CPC solar collector. In: 37th intersociety energy conversion engineering conference (IECEC), 2002, Paper No. 20150
35. CPC 6 XL INOX. www.rittersolar.de/english/index_e.htm. Accessed 2 Jan 2011
36. Evacuated tube collectors with German quality standard. www.linuo-ritter-international.com/products/evacuated-tube-collectors/. Accessed 10 Feb 2014
37. Thermosolar Group. www.thermosolar.de. Accessed 22 March 2014
38. Ritter Solar product catalog (2004) www.zeussolar.si/images/soncnikolektorji/SONCNI_KOLEKTORJI_KATALOG_2004_nem.pdf. Accessed 22 Jan 2010
39. Thermal products for water heating and space heating systems. www.consolar.co.uk/documents/Tubo%2012/Tubo%2012%20TDMA07_small_file.pdf. Accessed 6 Jan 2014
40. Macchi E (2013) The choice of working fluid: the most important step for a successful organic Rankine cycle (and an efficient turbine). In: 2nd international seminar on ORC power systems, Rotterdam
41. Pei G, Li J, Ji J (2010) Working fluid selection for low temperature solar thermal power generation with two-stage collectors and heat storage units. In: Manyala R (ed) Solar collectors and panels, theory and applications. ISBN:978-953-307-142-8 (InTech)
42. Liu B-T, Chien K-H, Wang C-C (2004) Effect of working fluids on organic Rankine cycle for waste heat recovery. Energy 29:1207–1217
43. Aljundi IH (2011) Effect of dry hydrocarbons and critical point temperature on the efficiencies of organic Rankine cycle. Renew Energy 36:1196–1202
44. Wei G, Yiwu W, Yuzhang W, Shaoqin S (2011) Heat recovery efficiency analysis of waste heat driven organic Rankine cycle. Acta Energiae Solaris Sinica 32:662–668
45. Maizza V, Maizza A (1996) Working fluids in non-steady flows for waste energy recovery systems. Appl Therm Eng 16:579–590
46. Chen H, Goswami DY, Stefanakos EK (2010) A review of thermodynamic cycles and working fluids for the conversion of low-grade heat. Renew Sustain Energy Rev 14:3059–3067
47. Li J, Pei G, Ji J (2010) Effect of working fluids on the efficiency of low-temperature solar-thermal-electric power generation system. Acta Energiae Solaris Sinica 31(581):587
48. Yamamoto T, Furuhata T, Arai N, Mori K (2001) Design and testing of the organic Rankine cycle. Energy 26:239–251
49. Huang Tzu-Chen (2001) Waste heat recovery of organic Rankine cycle using dry fluids. Energy Convers Manage 42:539–553
50. Hung TC, Wang SK, Kuo CH, Pei BS, Tsai KF (2010) A study of organic working fluids on system efficiency of an ORC using low-grade energy sources. Energy 35:1403–1411
51. Gozdur AB, Nowak W (2007) Comparative analysis of natural and synthetic refrigerants in application to low temperature Clausius-Rankine cycle. Energy 32:344–352

52. Abie Lakew A, Bolland O (2010) Working fluids for low-temperature heat source. Appl Therm Eng 30:1262–1268
53. Saleh B, Koglbauer G, Wendland M, Fischer J (2007) Working fluids for low-temperature organic Rankine cycles. Energy 32:1210–1221
54. Drescher U, Brueggemann D (2007) Fluid selection for the organic Rankine cycle (ORC) in biomass power and heat plants. Appl Therm Eng 27:223–228
55. Guo T, Wang HX, Zhang SJ (2011) Selection of working fluids for a novel low-temperature geothermally-powered ORC based cogeneration system. Energy Convers Manage 52:2384–2391
56. Saitoh TS, Kato J, Yamada N (2006) Advanced 3-D CPC solar collector for thermal electric system. Heat Transf-Asian Res 35:323–335
57. Wang EH, Zhang HG, Fan BY, Ouyang MG, Zhao Y, Mu QH (2013) Study of working fluid selection of organic Rankine cycle (ORC) for engine waste heat recovery. Energy 36:3406–3418
58. Tchanche BF, Lambrinos GR, Frangoudakis A, Papadakis G (2010) Exergy analysis of micro-organic Rankine power cycles for a small scale solar driven reverse osmosis desalination system. Appl Energy 87:1295–1306
59. Hettiarachchi HDM, Golubovic M, Wore WM, Ikegami Y (2007) Optimum design criteria for an organic Rankine cycle using low-temperature geothermal heat sources. Energy 32:1698–1706
60. Jian Xu, Dong Ao, Tao Li, Lijun Yu (2011) Working fluid selection of organic Rankine cycle (ORC) for low and medium grade heat source utilization. Energy Conserv Technol 29:204–210
61. Liu H, He Y, Cheng Z, Cui F (2010) Simulation of parabolic trough solar thermal generation with organic Rankine cycle. J Eng Thermophys 31(1631):1635
62. Wei DH, Lu XX, Lu Z, Gu JM (2007) Performance analysis and optimization of organic Rankine cycle (ORC) for waste heat recovery. Energy Convers Manage 48:1113–1119
63. Roy JP, Mishra MK, Misra A (2011) Performance analysis of an organic Rankine cycle with superheating under different heat source temperature conditions. Appl Energy 88:2995–3004
64. Dai YP, Wang JF, Gao L (2009) Parametric optimization and comparative study of organic Rankine cycle (ORC) for low grade waste heat recovery. Energy Convers Manage 50:576–582
65. Li W, Feng X, Yu LJ, Xu J (2011) Effects of evaporating temperature and internal heat exchanger on organic Rankine cycle. Appl Therm Eng 31:4014–4023
66. Xu RJ, He YL (2011) A vapor injector-based novel regenerative organic Rankine cycle. Appl Therm Eng 31:1238–1243
67. Bao JJ, Zhao L, Zhang WZ (2011) A novel auto-cascade low-temperature solar Rankine cycle system for power generation. Sol Energy 85:2710–2719
68. Badr O, Probert D, Callaghan PO (1986) Multi-vane expanders as prime movers in low-grade energy engines. Proc Inst Mech Eng Part A 200:117–125
69. James AM, Jon RJ, Cao J, Douglas KP, Richard NC (2009) Experimental testing of gerotor and scroll expanders used in, and energetic and exergetic modeling of, an organic Rankine cycle. J Energy Resoure-ASME 131(201):208
70. Lemort V, Quoilin S, Cuevas C, Lebrun J (2009) Testing and modeling a scroll expander integrated into an organic Rankine cycle. Appl Therm Eng 29:3094–3102
71. Quoilin S, Lemort V, Lebrun J (2010) Experimental study and modeling of an organic Rankine cycle using scroll expander. Appl Energy 87:1260–1268
72. Liu H, Qiu G, Shao Y, Daminabo F, Riffat SB (2010) Preliminary experimental investigations of a biomass-fired micro-scale CHP with organic Rankine cycle. Int J Low-Carbon Technol 2(81):87
73. Riffat SB, Zhao X (2004a) A novel hybrid heat pipe solar collector/CHPsystem—Part 1: System design and construction. Renew Energy 29:2217–2233

74. Riffat SB, Zhao X (2004b) A novel hybrid heat-pipe solar collector/CHP system—Part II: Theoretical and experimental investigations. Renew Energy 29:1965–1990
75. Kane M, Larrain D, Favrat D, Allani Y (2003) Small hybrid solar power system. Energy 28:1427–1443
76. Peterson RB, Wang H, Herron T (2008) Performance of small-scale regenerative Rankine power cycle employing a scroll expander. Proc Inst Mech Eng Part A: J Power Energy 222 (271):282
77. Manolakos D, Kosmadakis G, Kyritsis S, Papadakis G (2009) Identification of behaviour and evaluation of performance of small scale, low-temperature organic Rankine cycle system coupled with a RO desalination unit. Energy 34:767–774
78. Pei G, Li J, Ji J (2010) Analysis of low temperature solar thermal electric generation using regenerative organic Rankine cycle. Appl Therm Eng 30:998–1004
79. Li J, Pei G, Ji J (2010) Optimization of low temperature solar thermal electric generation with organic Rankine cycle in different areas. Appl Energy 87:3355–3365
80. Al-Sulaiman FA, Dincer I, Hamdullahpur F (2012) Energy and exergy analyses of a biomass trigeneration system using an organic Rankine cycle. Energy 45:975–985
81. Wang H, Peterson R, Herron T (2011) Design study of configurations on system COP for a combined ORC (organic Rankine cycle) and VCC (vapor compression cycle). Energy 36:4809–4820
82. Bu XB, Li HS, Wang LB (2013) Performance analysis and working fluids selection of solar powered organic Rankine-vapor compression ice maker. Sol Energy 95:271–278
83. Jradi M, Riffat S (2014) Experimental investigation of a biomass-fuelled micro-scale tri-generation system with an organic Rankine cycle and liquid desiccant cooling unit. Energy 71:80–93
84. Infinity Turbine LLC. www.infinityturbine.com/ORC/ORC_Waste_Heat_Turbine.html. Accessed 16 Oct 2010
85. Mills D (2004) Advances in solar thermal electric technology. Sol Energy 76:19–31
86. Buttinger F, Beikircher T, Proll M, Scholkopf W (2010) Development of a new flat stationary evacuated CPC-collector for process heat applications. Sol Energy 84:1166–1174
87. Jadhav AS, Gudekar AS, Patil RG, Kale DM, Panse SV, Joshi JB (2013) Performance analysis of a novel and cost effective CPC system. Energy Convers Manage 66:56–65
88. Sagade AA, Shinde NN, Patil PS (2014) Effect of receiver temperature on performance evaluation of silver coated selective surface compound parabolic reflector with top glass cover. Energy Procedia 48:212–222
89. Liu Z, Tao G, Lu L, Wang Q (2014) A novel all-glass evacuated tubular solar steam generator with simplified CPC. Energy Convers Manage 86:175–185
90. Fernández A, Dieste JA (2013) Low and medium temperature solar thermal collector based in innovative materials and improved heat exchange performance. Energy Convers Manage 75:118–129
91. Gang P, Guiqiang L, Xi Z, Jie J, Yuehong S (2012) Experimental study and exergetic analysis of a CPC-type solar water heater system using higher-temperature circulation in winter. Sol Energy 86:1280–1286
92. Li X, Dai YJ, Li Y, Wang RZ (2013) Comparative study on two novel intermediate temperature CPC solar collectors with the U-shape evacuated tubular absorber. Sol Energy 93:220–234
93. Yoshioka K, Suzuki A, Saitoh T (1999) Performance evaluation of two-dimensional compound elliptic lens concentrators using a yearly distributed insolation model. Sol Energy Mater Sol Cells 57:9–19
94. Mallick TK, Eames PC, Hyde TJ, Norton B (2006) Non-concentrating and asymmetric compound parabolic concentrating building facade integrated photovoltaics: an experimental comparison. Sol Energy 80:834–849
95. Garboushian V, Roubideaux D, Yoon S (1997) Integrated high-concentration PV Nearterm alternative for low-cost large-scale solar electric power. Sol Energy Mater Sol Cells 47:315–323

96. Marie Curie Actions Incoming International Fellowships, Lens-walled CPC, FP7-PEOPLE-2009-IIF
97. Su Y, Riffat SB, Pei G (2012) Comparative study on annual solar energy collection of a novel lens-walled compound parabolic concentrator (lens-walled CPC). Sustain Cities Soc 4:35–40
98. Su Y, Pei G, Riffat SB, Huang H (2012) A novel lens-walled compound parabolic concentrator for photovoltaic applications. J Sol Energy Eng 124:021010–021017
99. Guiqiang L, Gang P, Yuehong S, jie J, Riffat SB (2013) Experiment and simulation study on the flux distribution of lens-walled compound parabolic concentrator compared with mirror compound parabolic concentrator. Energy 58:398–403
100. Delgado-Torres AM, García-Rodríguez L (2010) Analysis and optimization of the low temperature solar organic Rankine cycle (ORC). Energy Convers Manage 51:2846–2856
101. Delgado-Torres AM, García-Rodríguez L (2007a) Preliminary assessment of solar organic Rankine cycles for driving a desalination system. Desalination 216:252−275
102. Li C, Kosmadakis G, Manolakos D, Stefanakos E, Papadakis G, Goswami DY (2013) Performance investigation of concentrating solar collectors coupled with a transcritical organic Rankine cycle for power and seawater desalination cogeneration. Desalination 318:107–117
103. Nafey AS, Sharaf MA, García-Rodríguez L (2010) Thermo-economic analysis of a combined solar organic Rankine cycle-reverse osmosis desalination process with different energy recovery configurations. Desalination 261:138–147
104. Kosmadakis G, Manolakos D, Kyritsis S, Papadakis G (2009) Economic assessment of a two-stage solar organic Rankine cycle for reverse osmosis desalination. Renew Energy 34:1579–1586
105. Delgado-Torres AM, García-Rodríguez L (2007b) Double cascade organic Rankine cycle for solar-driven reverse osmosis desalination. Desalination 216:306−313
106. Kosmadakis G, Manolakos D, Papadakis G (2010) Parametric theoretical study of a two-stage solar organic Rankine cycle for RO desalination. Renew Energy 35:989–996
107. Delgado-Torres Agustín M, García-Rodríguez Lourdes (2012) Design recommendations for solar organic Rankine cycle (ORC)–powered reverse osmosis (RO) desalination. Renew Sustain Energy Rev 16:44–53
108. Manolakos D, Kosmadakis G, Kyritsis S, Papadakis G (2009) On site experimental evaluation of a low-temperature solar organic Rankine cycle system for RO desalination. Sol Energy 83:646–656
109. Nafey AS, Sharaf MA (2010) Combined solar organic Rankine cycle with reverse osmosis desalination process: energy, exergy, and cost evaluations. Renew Energy 35:2571–2580
110. Quoilin S, Orosz M, Hemond H, Lemort V (2011) Performance and design optimization of a low-cost solar organic Rankine cycle for remote power generation. Sol Energy 85:955–966
111. Wang XD, Zhao L, Wang JL, Zhang WZ, Zhao XZ, Wu W (2010) Performance evaluation of a low-temperature solar Rankine cycle system utilizing R245fa. Sol Energy 84:353–364
112. He YL, Mei DH, Tao WQ, Yang WW, Liu HL (2012) Simulation of the parabolic trough solar energy generation system with organic Rankine cycle. Appl Energy 97:630–641
113. Wang J, Yan Z, Zhao P, Dai Y (2014) Off-design performance analysis of a solar-powered organic Rankine cycle. Energy Convers Manage 80:150–157
114. Dumont O, Declaye S, Quoilin S, Lemort V (2013) Design, modelling and experimentation of a small-scale solar ORC. In: 2nd international seminar on ORC power systems, Rotterdam
115. Dickes R, Orosz MS, Hemond HF (2013) Non-constant wall thickness scroll expander investigation for micro solar ORC plant. In: 2nd international seminar on ORC power systems, Rotterdam
116. Zhang Y-Q, Guo H, Wu Y-T, Xia G-D, Ma C-F (2013) Performance study on distributed trough solar power system based on single screw expander and organic Rankine cycle. In: 2nd international seminar on ORC power systems, Rotterdam
117. Colonna P, Bahamonde S (2013) Solar ORC turbogenerator for green-energy buildings. In: 2nd international seminar on ORC power systems, Rotterdam

Chapter 2
Structural Optimization of the ORC-Based Solar Thermal Power System

Owing to the low temperature difference between the hot and the cold sides, the thermal efficiency of solar ORC is much lower than that of fossil-fired steam plants. The efficiency restriction by thermodynamic laws is extremely critical for small-scale solar ORCs, in which the expander, generator, pump, etc. are less efficient than those in the steam Rankine cycle. For the sake of an acceptable power efficiency, it is necessary to minimize the thermodynamic irreversibility of the solar ORC system. And structural optimization is performed in this chapter. The total losses of the ORC can be divided into four parts, i.e. losses in the heating, expansion, cooling and pressurization processes, with corresponding devices of evaporator, expander, condenser and pump. Exergy losses in the cooling process are related to the degree of superheat of exhaust leaving the expander. They are lessened when an internal heat exchanger is employed. Exergy losses in the expansion process are inevitable because the expander efficiency is usually lower than 0.85. There is great technical difficulty in achieving higher expander efficiency. Exergy losses in the heating and pressurization processes, on the other hand, are of special interest.

The potential of exergy reduction in the heating process is large. The irreversibility in the conventional single-stage evaporator (SSE) is large because of the great temperature difference between the HTF and the ORC fluid, as illustrated in Fig. 2.1. The vertical axis is the temperature, and the horizontal is the heat transferred from HTF to ORC fluid. Adverse current heat exchanger is exemplified. Given the pinch point temperature (ΔT_p) and the inlet and outlet conditions of the ORC fluid, a high mass flow of HTF is accompanied with a high HTF outlet temperature while a low one results in a high HTF inlet temperature. It is difficult to lower the heat transfer irreversibility in the SSE. And the associated exergy losses amount to about 40 % of the total ORC exergy destruction [1]. To settle this problem, two structures are proposed in the solar ORC: collector for direct vapor generation (CDVG), and collector integrated PV module. The SSE is then replaced.

Aside from the evaporator, the pump may play an important role in small scale ORCs. To get a close view of this, a comparison between the ORC pump and the pump used in the steam Rankine cycle can be made.

J. Li, *Structural Optimization and Experimental Investigation of the Organic Rankine Cycle for Solar Thermal Power Generation*, Springer Theses,
DOI 10.1007/978-3-662-45623-1_2

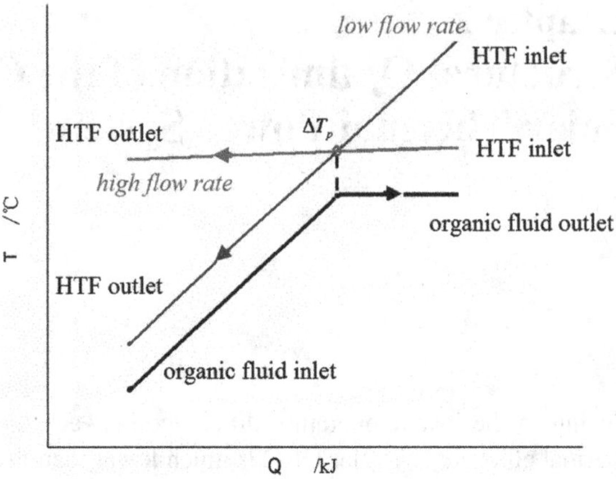

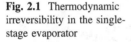

Fig. 2.1 Thermodynamic irreversibility in the single-stage evaporator

First, the overall pumping efficiency in a small ORC is supposed to be much lower. There are several reasons: (1) The self-efficiency of the ORC pump, which is defined as the ratio of the theoretical power requirement to the practical power delivered to the pump shaft [2], is low. Volumetric and centrifugal pumps are currently widely used to drive refrigerants. The leakage loss due to a gap between the bottom/top plate and the scrolls, or between the flanks of the scrolls, or between the back surface of the impeller hub plate and the casing, and the hydro-mechanical loss attributed by the friction between the organic fluid and the walls, acceleration and retardation of the fluid and the change of the flow direction generally get more appreciable as the dimension of the pump is reduced. Though in steam Rankine cycle it is easy for a centrifugal pump to operate at an efficiency of up to 90 %, small pumps have a much lower efficiency, which is commonly 45–70 % in kW range [3–5]. (2) The loss in the pump is just part of total loss in the pumping process. The shaft power of a pump is transmitted by a prime mover rather than electricity. The mover can be a diesel, gasoline engine or motor. Among current drivers for pumps, the induction motor is most widely used owing to its simple construction [6]. The motor is equipped with a pump and coupling, and its efficiency is expressed by the ratio of the shaft output power to electrical input power [7]. Similarly to the pump, the motor friction losses, bearing losses, electrical and magnetic losses become more significant at smaller sizes. An efficiency from 50 to 75 % is normal for a small motor. Working together with the motor, a frequency converter is favorably used to control the pump in an energy-saving way. Via a conversion of a current with one frequency to a current with another frequency, the converter controls the speed and the torque of the motor, resulting in attractive energy conservation when the power requirement of the ORC pump is not constant. Nevertheless, power loss exists in this conversion. In regard to these inefficiencies, the "wire-to-liquid" efficiency in the pumping process can be as low as 30 % even at design conditions (efficiency of 50 % for pump, 65 % for motor and 88 % for converter). (3) The ORC will suffer from even lower "wire-to-liquid" efficiency at off-design conditions. Cogeneration or tri-generation of the ORC leads to energy savings compared with stand-alone systems. One challenge is that consumers' demands on heat and power follow a season cycle, and thus a flexible trade-off between the needs for heat and power should be permitted. Heat source properties of the ORC (e.g. the intensity of solar irradiation, the heating value of biomass) also fluctuate. Variable operation is necessitated for small scale ORC systems [8]. The off-design operation will lead to further degradation in performance of the pump, motor and converter. And a global pumping efficiency of around 25 % can be expected in the practical operation (efficiency of 45 % for pump, 60 % for motor and 85 % for converter).

Second, the negative power associated with the ORC pump is much more appreciable. The power consumed by the pump is only 1–2 % of the work generated by the expander in the steam Rankine cycle [9]. In approximating the net cycle output, the power for water pumping is usually negligible. But this neglect will be no more reasonable for the ORC. The ratio of the electricity consumed in the pumping process and that generated on different conditions is presented in Table 2.1. The efficiencies of the expander and generator are 0.75 and 0.80, respectively. The efficiencies of the pump, motor and converter are 0.5, 0.6, and 0.85. In most cases, the electricity consumed and that generated have the same order of magnitude. For each of the fluid, the ratio increases with the increment in the evaporation and condensation temperatures. Given the temperatures, the ratio gets larger when a fluid of lower boiling point is selected. It is clear the negative work in the small ORC is significant. The net cycle efficiency can even be zero or negative depending on the electricity consumption in pumping [10]. Among the components of the conventional ORC, the pump tends to cause the most critical decline in the thermal efficiency [11].

Third, the technical requirement of the ORC pump is stricter. A common problem associated with the water pump is the leakage. Unlike water, most of ORC fluids are either toxic, flammable, ozone layer depleting or have global warming impact. And the fluids are much more expensive than water. The ORC pump must be very well sealed to prevent loss of the fluid to the atmosphere. Any inward

Table 2.1 Ratio of electricity consumed and that generated in the ORC, unit: %

Working fluid	Condensation temperature (°C)	Evaporation temperature (°C)					
		100	110	120	130	140	150
R245fa	30	15.19	17.42	20.01	23.00	26.71	32.27
	40	17.33	19.75	22.74	26.16	30.40	36.60
	50	20.00	22.82	26.05	29.96	34.78	42.12
	60	22.46	25.90	29.43	33.96	39.52	47.81
R123	30	9.26	10.59	12.05	13.78	15.63	17.91
	40	10.45	11.89	13.65	15.51	17.66	20.11
	50	12.08	13.60	15.48	17.68	19.97	22.77
	60	14.09	15.94	17.90	20.22	22.87	25.87

Note The fluid at the expander and pump inlet is saturated vapor and liquid, respectively

leakage of air into the pump should be also interrupted. And great attention must be paid to the compatibility of the sealing material and the organic fluid, which limits the usage of seals. For example, as a widely used hydrocarbon elastomer Buna-N has excellent resistance to petroleum-based oils, water, silicone greases, hydraulic fluids and alcohols, but it is not compatible with R123.

Fourth, cavitation in the ORC pump is more easily facilitated, which is attributed to the much higher saturation vapor pressure of organic fluids at ambient temperature. The pump's inefficiency and the fluid's low heat capacity will result in significant temperature increment in the pumping process. In a lot of theoretical analysis, the organic fluid at the pump inlet is assumed to be saturated. However, when considering the temperature increment during the irreversible pressurization, the fluid needs to be sub-cooled in practical operation. Measures have to be taken to increase the inlet pressure in meters head for the pump, such as enlarging the height difference between the pump and the reservoir.

Fifth, the cost per kW of the ORC pump is higher, which results from the strict technical requirement and the low power of the ORC pump. As the heat exchanger area and material cost are almost linearly proportional with the load, the pump will amount to a larger proportion of the total cost of the ORC at smaller power [12].

To address the above problems associated with the ORC pump, the osmosis-driven solar ORC is proposed. The pressurization process in the ORC no longer takes place in the pump.

Above all, this chapter is focused on three types of systems: solar ORC with CDVG, solar ORC with PV module and osmosis-driven solar ORC.

2.1 System Description

2.1.1 Solar ORC with CDVG

In the previous work by the author, two-stage heat exchangers have been proposed to reduce the heat transfer irreversibility in the SSE, as shown in Fig. 2.2. The first-stage heat exchanger is a preheater through which the organic fluid can be heated from sub-cooled to saturated liquid condition. The second-stage heat exchanger is a boiler. The two-stage heat exchangers are connected with the CPC collectors independently. The mass flow rate of HTF in the first-stage heat exchanger (E1) is lower than that in the second-stage heat exchanger (E2). According to the T-Q curve, the two-stage heat exchangers can offer a relatively low temperature difference between the organic fluid and HTF. The average operating temperature of the collectors is diminished and thus the overall heat collection efficiency is improved. Notably, in the presence of HTF, more pumps and heat exchangers are required to transfer solar heat to the ORC.

By employing the CDVG, the HTF can be eliminated. The feasibility of CDVG has been demonstrated by using the working fluid of water. A low-cost all-glass evacuated tubular solar steam generator with simplified CPC is capable to produce

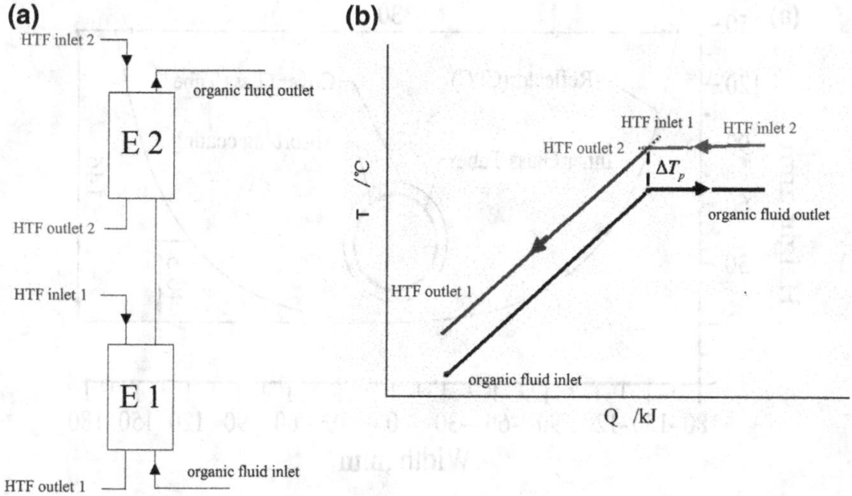

Fig. 2.2 Two-stage heat exchangers: **a** structure; **b** T-Q curves

steam exceeding 200 °C with pressure ranging from 100 to 550 kPa, as introduced by Liu et al. [13]. The generator consists of 60 collecting units and each unit is comprised of an all-glass evacuated tube, a simplified CPC reflector, and a metal concentric annular tube inserted inside the evacuated tube. The mixture of conduction oil and graphite powder is filled in the annular space between the inner glass tube and the copper concentric tube for the sake of efficient heat transfer from the selective absorbing layer to the water/steam in the concentric tube, as shown in Fig. 2.3. The collector combining the external CPC and the U-type evacuated tube has also been demonstrated [14, 15]. The cross-sectional structure of the module is shown in Fig. 2.4. The space between the envelope and the inner tube is a vacuum, and the inner tube is plated with the aluminum fin used as the heat transfer component. The heat absorbed by the inner tube is conducted to the U-type pipe by the fin. This kind of collector can handle high pressure without leakage of the working fluid. An efficiency of 46 % at the operating temperature of 200 °C can be achieved on use of 6X CPC collectors. Though CPC collectors operating with organic fluids are rarely investigated, they are expected to be competent in direct organic vapor generation on the use of the metal tube/pipe technologies.

2.1.1.1 Structure and Fundamental

There can be two ways to realize direct vapor generation in solar ORC, which are characterized by one-stage collector and heat storage unit, and two-stage collectors and heat storage units, respectively. The former is illustrated in Fig. 2.5. The system consists of CPC collectors, pumps and a fluid storage tank with phase change material (PCM), expander (T), generator (G), regenerator (R) and condenser. The CPC

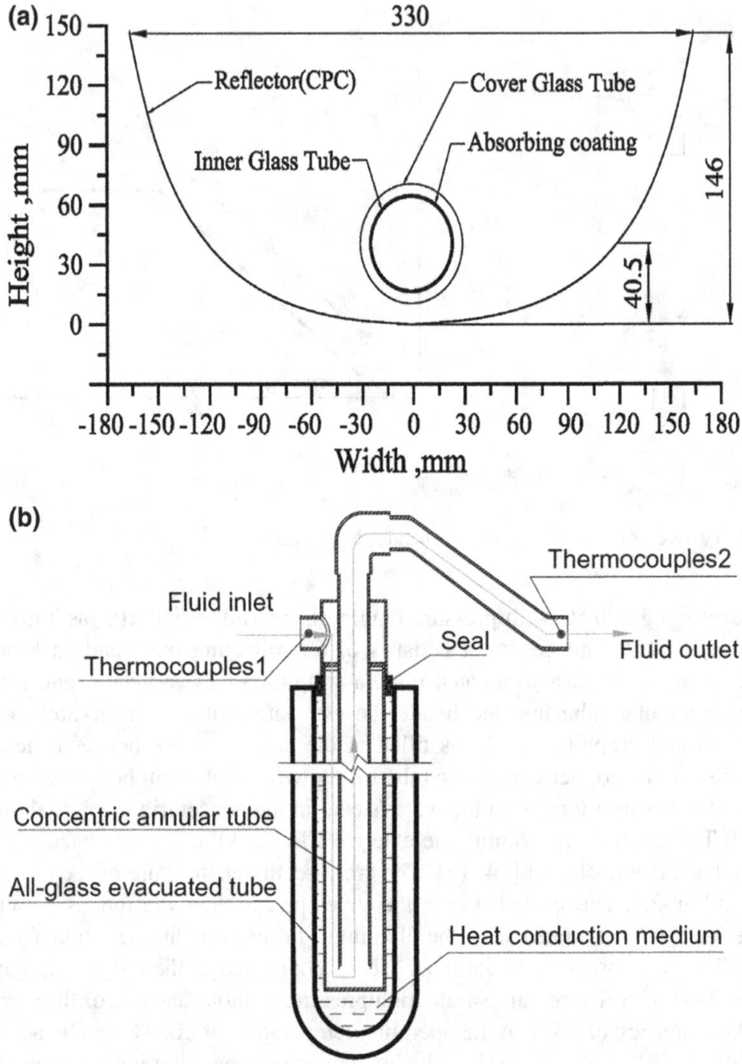

Fig. 2.3 Cross-sections of a direct steam generator: **a** the collector; **b** the inner tube. Reprinted from Ref. [13], Copyright 2010, with permission from Elsevier

collectors serve as the direct vapor generator. In contrast to the traditional solar Rankine system, there is an organic fluid storage tank with PCM at the inlet of the expander. In the practical operation there can be three modes:

(I) The system needs to generate electricity and solar radiation is available. In this mode, valves 1, 2 and 3 are open. Pump 1 is running. The organic fluid is heated and vaporized through the collectors under high pressure. The vapor flows into the expander, exporting power in the process due to the pressure

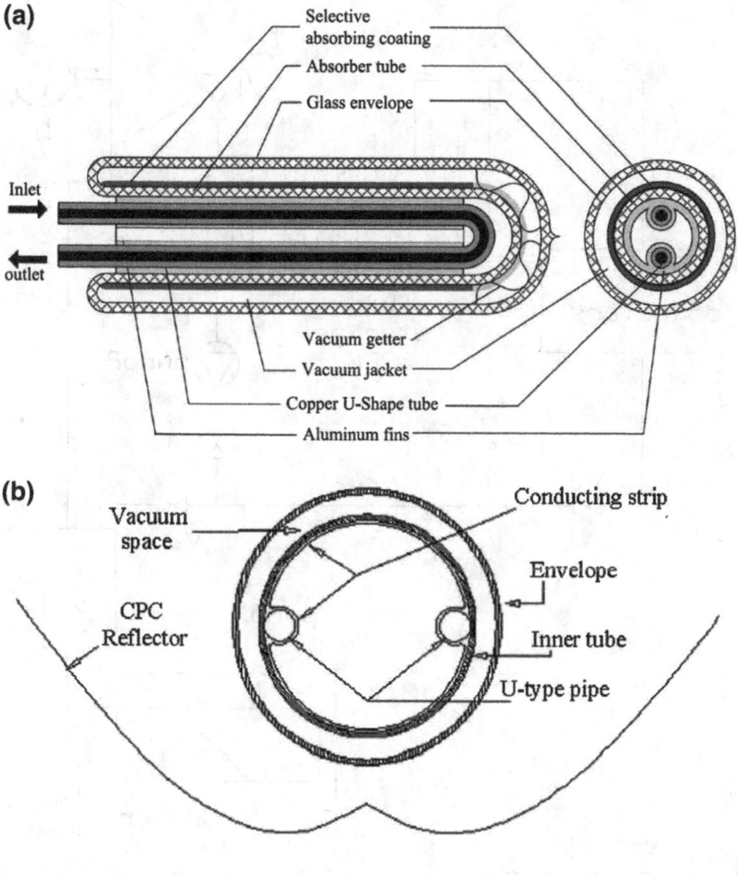

Fig. 2.4 Cross and profile sections of a U-type CPC. Reprinted from Refs. [14, 15], Copyright 2010, with permission from Elsevier

drop. The outlet vapor is cooled down in the regenerator and condensed to a liquid state in the condenser. The liquid is pressurized by pump 1 and warmed in the regenerator. The organic fluid is then sent back to the collectors and circulates. In case the irradiation is too strong, valve 4 would be open and pump 2 would run to prevent the organic fluid from being superheated in the collectors and part of the solar energy is stored.

(II) The system does not need to generate electricity but irradiation is well. Valves 3 and 4 are open. Pump 2 is running. In this mode, the solar heat is transferred to the PCM by the organic fluid. Heat storage is in process.

(III) The system needs to generate electricity but irradiation is very weak or unavailable. Valves 1, 2 and 5 are open and pump 1 is running. Heat is released from the PCM and converted into power by the ORC.

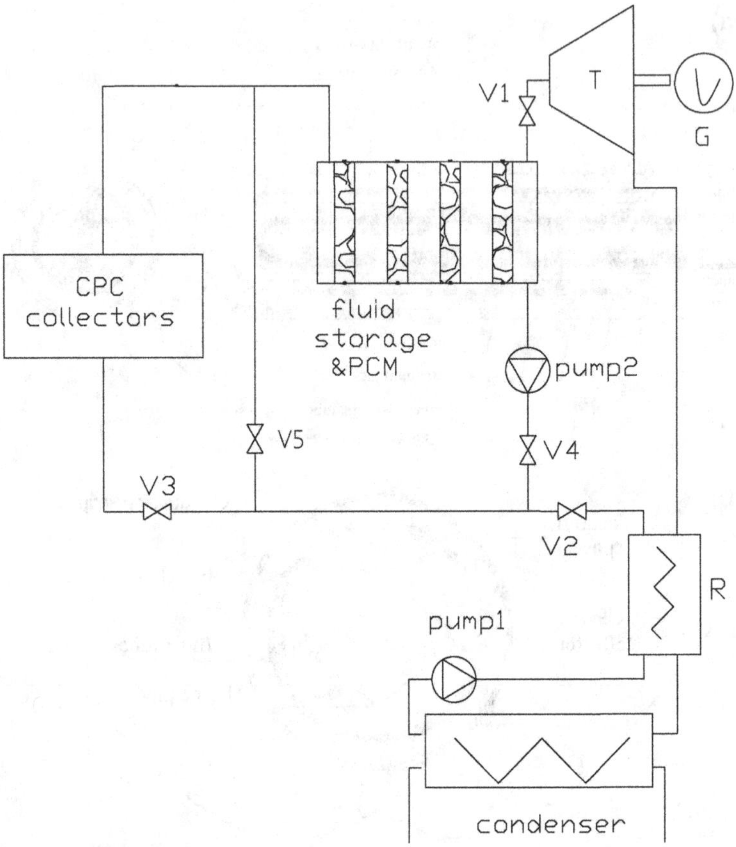

Fig. 2.5 Solar ORC with CDVG using one-stage collector and heat storage unit

Mode I presents the simultaneous processes of heat collection and power generation while Mode II or Mode III is the independent process of heat collection or power conversion.

The solar ORC with two-stage collectors and heat storage units is illustrated in Fig. 2.6. There are also three basic modes. In Mode I, valves 1, 2, 3 and 6 are open. Pump 1 is running. The organic fluid is preheated in the FPC collectors and then is vaporized in the CPC collectors. In Mode II, valves 3, 4 and 5 are open. Pumps 2 and 3 are running. The dashed lines are for heat storage. Solar heat collected by FPCs and CPCs is transferred to PCM (1) and PCM (2), relatively. The melting point of PCM (1) is lower than that of PCM (2). In Mode III, valves 1, 2 and 5 are open. Pump 1 is running. The organic fluid first absorbs the heat released by PCM (1) and then is vaporized in the tank filled with PCM (2).

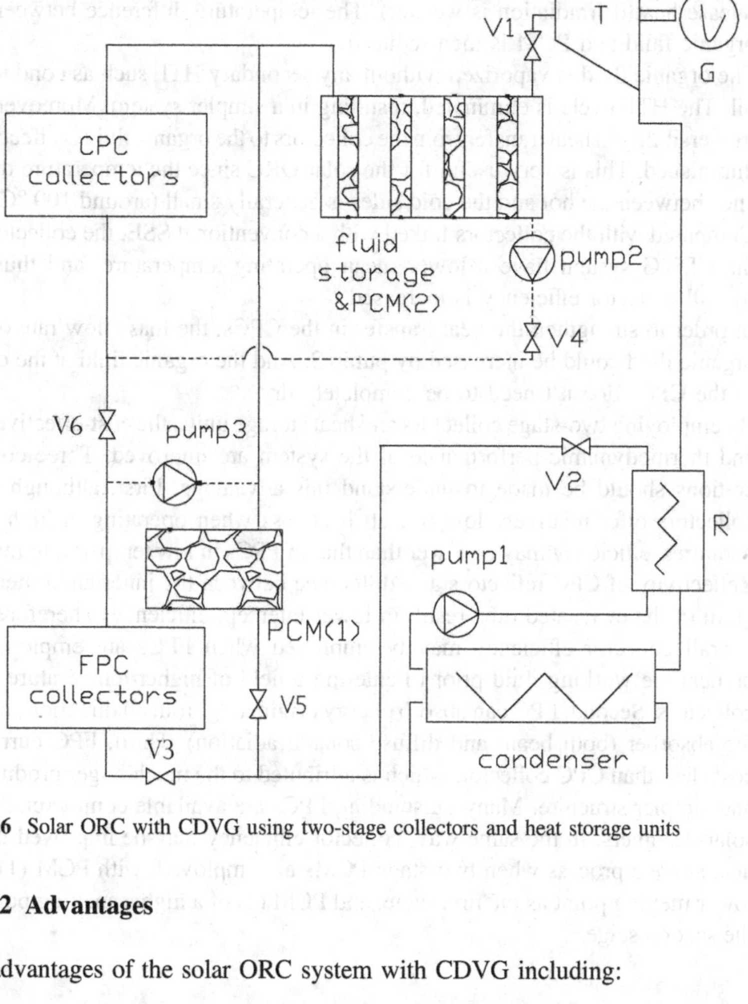

Fig. 2.6 Solar ORC with CDVG using two-stage collectors and heat storage units

2.1.1.2 Advantages

The advantages of the solar ORC system with CDVG including:

(1) By placing a fluid storage tank filled with PCM at the expander inlet, the stability of the ORC subsystem can be guaranteed. Owing to the storage unit, the temperature of the organic fluid in the tank is close to the melting point of the PCM. Since the fluid is at a binary-phase state, the pressure in the tank and hence the flow rate through the expander are almost constant. If the vapor feed-in rate is smaller than that through the expander, the pressure will become lower than the saturated value, and the liquid in the tank will be vaporized. Similarly, a larger feed-in rate will be accompanied by a higher pressure than the saturated, and the residual vapor will be condensed.

(2) Without any complicated control device, the processes of heat storage or heat release can take place while electricity is being generated. The organic fluid is

able to exchange energy with the PCM in the storage tank at a relatively low heat flux (store heat if irradiation is stronger than that on normal condition or release heat if irradiation is weaker). The temperature difference between the organic fluid and PCM is then reduced.

(3) The organic fluid is vaporized without any secondary HTF such as conduction oil. The HTF cycle is eliminated, resulting in a simpler system. Moreover, the irreversibility in heat transfer from the collectors to the organic fluid is effectively diminished. This is very useful for the solar ORC since the temperature difference between the hot and the cold sides is generally small (around 100 °C).

(4) Compared with the collectors linked with a conventional SSE, the collectors in the CDVG system have a lower mean operating temperature, and thus the overall collector efficiency is increased.

(5) In order to strengthen the heat transfer in the CPCs, the mass flow rate of the organic fluid could be increased by pump 2. And the organic fluid at the outlet of the CPCs doesn't need to be completely dry.

(6) By employing two-stage collectors and heat storage units, the cost-effectiveness and thermodynamic performance of the system are improved. Three considerations should be made to understand this advantage. First, although CPC collectors offer relatively low overall heat loss when operating at high temperatures, efficiency may be lower than that of FPCs in low temperature ranges. Reflectivity of CPC reflectors and difference between the inner and outer diagram of the evacuated tube result in lower intercept efficiency. Therefore, the overall collector efficiency may be improved when FPCs are employed to preheat the working fluid prior to entering a field of higher-temperature CPC collectors. Second, FPC can absorb energy originating from all directions above the absorber (both beam and diffuse solar irradiation). Third, FPC currently costs less than CPC collector, which is attributed to the much larger production and simpler structure. Many outstanding FPCs are available commercially for solar designers. In the same way, collector efficiency may be improved in the heat storage process when two-stage PCMs are employed, with PCM (1) of a lower melting point as the first stage, and PCM (2) of a higher melting point as the second stage.

2.1.2 Solar ORC with PV Module

2.1.2.1 Background

Photovoltaic (PV) is the simplest way to generate electricity from solar irradiation. Both the power efficiency and cost-effectiveness of the cells have been improved significantly in the past 15 years. One of the challenges associated with solar PV system is the issue of intermittency. Currently lots of the PV systems are connected to the electric grid without storage. Solar cells will generate electricity steadily throughout the day as long as the sun is shining. But when the sky is partly cloudy, solar power output from an entire region can fluctuate unpredictably. To mitigate the impact of this phenomenon, energy storage is a vital component of a more resilient, reliable and efficient electric grid. The grid's evolution toward more distributed energy systems and the incorporation of electric vehicles and plug-in hybrids is contributing to the growing interest in grid storage. And energy storage will help lower consumer costs by saving low-cost power for peak times and making renewable energy available when it's needed the most, not just when the sun is shining. Aside from the grid-tied system, the off-grid PV system is preferable in remote areas for the sake of a silent, emission-free energy source, avoidance of the high cost of extending a utility line, and independence of homemade energy production. A backup energy source is essential for the stand-alone PV system.

On the other hand, the deployment of battery storage will dramatically increase the cost of entire PV system. Take the lithium ion battery for example. It is a commonly used type of rechargeable battery with a fast-growing market, and has the advantages of light weight for a given capacity, high open-circuit voltage, low self-discharge rate (about 1.5 % per month), and reduced toxic landfill. The cost of lithium ion battery is about 4,500 RMB/kWh in China. The current cost of a PV system without storage is about 8,000 RMB/kW. When adding a battery storage of 5 h, the total cost becomes 30,500 RMB/kW, which is about 4 times as the initial [16]. The lead-acid battery is another widely used rechargeable battery and is relatively cheaper. But it has the problems of sulfation, risk of exploding under certain conditions, and environmental issues. Sulfation occurs when the battery becomes old or stays discharged for long enough. Some lead compounds contained in the battery are extremely toxic. They can cause damage especially to children who are still developing. Besides, the battery requires replacement every few years.

Compared with battery storage, heat storage is much cheaper. A lots of PCMs can be applied in the low-medium temperature solar thermal power system. And some information is presented in Table 2.2. With a capacity of 5 h storage and a solar thermal electric efficiency of 8 %, the cost associated with the thermal storage is about 1,300 RMB/kW when using $MgCl_2 \cdot 6H_2O$, which is one order of magnitude lower than that with battery storage.

To combine the advantages of the PV and thermal storage, the solar ORC with PV module is proposed.

Table 2.2 Property and cost of some PCMs

	Melting point (°C)	Latent heat (kJ/kg)	Conductivity (W m^{-1} K^{-1})	Density (kg m^{-3})	Cost (RBM/ton)
Paraffin (PNW106)	106 ± 2	80	0.65	1,200	4,000
$MgCl_2 \cdot 6H_2O$	117	168.6	0.57	1,569	1,000
Erythritol	120	339.8	0.73	1,450	12,000
D-mannitol	167 ± 1	316.4	–	1,520	/
Dulcitol	189	351.8	–	1,470	/

Note The cost of the PCMs is achieved from website of Chinese chemistry product [17]

2.1.2.2 Structure and Fundamental

Figure 2.7 is the scheme of the solar ORC with PV module. The system mainly consists of the PV-CPC modules and the ORC subsystem. Figure 2.8 shows the cross section of an alternative hybrid PV-CPC module. The organic fluid is vaporized in the tube of PV-CPC module. The PV cells are packed between two transparent layers, with intermediate layer of high temperature silicone elastomer in between, as shown in Fig. 2.9. The whole lot of PV cells and transparent layers is pasted over a black absorber. Metallic panel-groove is put underneath the absorber to improve heat conduction and the working fluid tube is sealed between the panel-groove and the absorber. The adhesion process should be under precise pressure control to ensure good quality thermal conductance [18].

The hybrid system generates electricity directly by the PV cells. And the waste heat from the cells is used to drive the ORC. From the viewpoint of thermal power

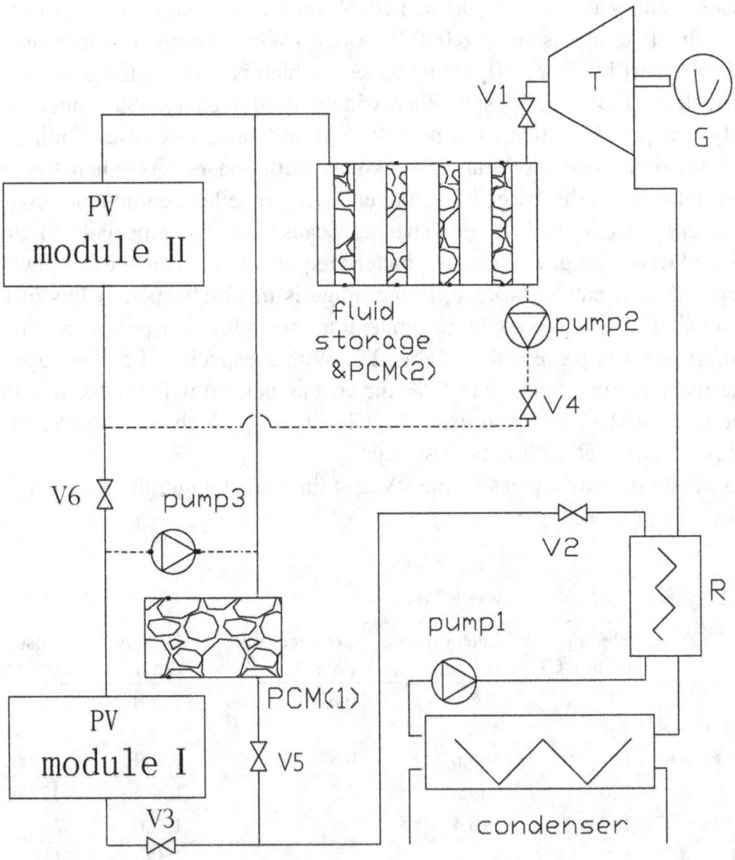

Fig. 2.7 Solar ORC with PV modules

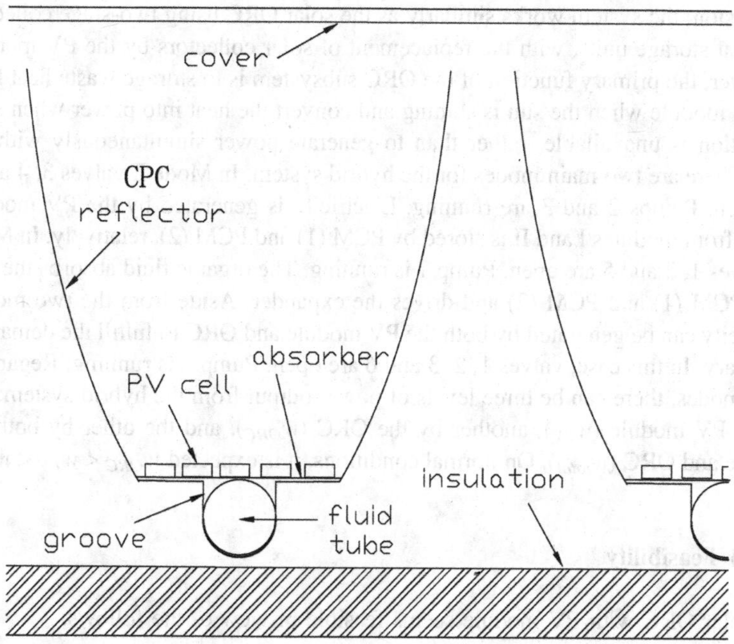

Fig. 2.8 Cross section of the PV module. Reprinted from Ref. [18], Copyright 2010, with permission from Oxford University Press

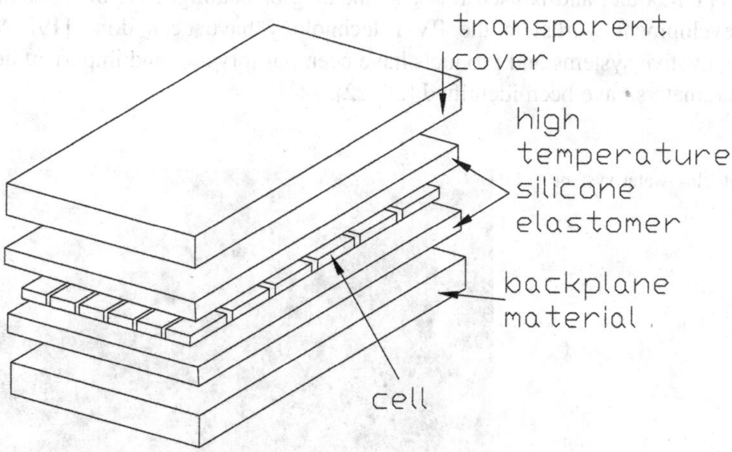

Fig. 2.9 Layout of the PV cell. Reprinted from Ref. [18], Copyright 2010, with permission from Oxford University Press

conversion, the system works similarly as the solar ORC using two-stage collectors and heat storage units, with the replacement of solar collectors by the PV module. However, the primary function of the ORC subsystem is to storage waste heat from the PV module when the sun is shining and convert the heat into power when solar irradiation is unavailable, rather than to generate power simultaneously with PV cells. There are two main modes for the hybrid system. In Mode I, valves 3, 4 and 5 are open. Pumps 2 and 3 are running. Electricity is generated by the PV module. Waste from modules I and II is stored by PCM (1) and PCM (2), relatively. In Mode II, valves 1, 2 and 5 are open. Pump 1 is running. The organic fluid absorbs the heat from PCM (1) and PCM (2) and drives the expander. Aside from the two modes, electricity can be generated by both the PV module and ORC to fulfill the demand if necessary. In this case, valves 1, 2, 3 and 6 are open. Pump 1 is running. Regarding these modes, there can be three levels of power output from the hybrid system: one by the PV module (w_{PV}), another by the ORC (w_{ORC}), and the other by both PV module and ORC (w_{total}). On normal conditions, it is expected $w_{ORC} < w_{PV} < w_{total}$.

2.1.2.3 Feasibility

The solar ORC with PV module seems feasible in view of the follows:

(1) Hybrid photovoltaic/thermal (PV/T) technology at low temperature has been well demonstrated. The PV/T modules produce electricity and heat simultaneously, and thus the overall efficiency in solar energy utilization is significantly increased. Some PV/T-air and PV/T-water systems are presented in Figs. 2.10 and 2.11. The air or water takes away the waste heat when flowing through the PV/T module, and is used for space heating or bathing. Lots of research and development works on the PV/T technology have been done [19]. Many innovative systems and products have been put forward, and important design parameters have been identified [20–22].

Fig. 2.10 PV-water system

Fig. 2.11 PV-air system (Trombe wall)

(2) The research in PV/T system operating at medium temperature is also active. A system of this kind is shown in Fig. 2.12. The applications include solar cooling [23, 24], solar heating [25], and desalination [26]. Some PV cells such as InGaP/InGaAs/Ge triple-junction and Schottky barrier have been experimentally investigated [27, 28]. These cells can tolerate high temperature up to 170 °C, while maintaining reasonable electric conversion efficiency. Besides, solar cells are available for commercial missions such as communications satellites, weather satellites, and earth-orbiting missions, for which the PV temperature fluctuation may be from −200 to 200 °C [29].

(3) The hybrid solar ORC/PV system may work at temperature around 100 °C, the PV efficiency loss will not be drastic. And $\alpha - Si$ cell would be economically and technologically suitable for this application. $\alpha - Si$ cell has relatively low temperature coefficients of maximum power generated, which is about 0.21 %/°C. And it is relatively cheaper. By using high temperature silicone elastomer $\alpha - Si$ cell could work at high temperature and remain efficient, e.g. about 85 % of the standard efficiency at 100 °C. Some $\alpha - Si$ cells available on the market have already complied with qualification including high-temperature thermal cycle and damp-heat testing [30].

(4) The overall electricity efficiency of the hybrid system is expected to be higher than that of a conventional stand-alone PV system [31]. The heat storage helps improve the device utilization ratio, and leads to a shorter payback time.

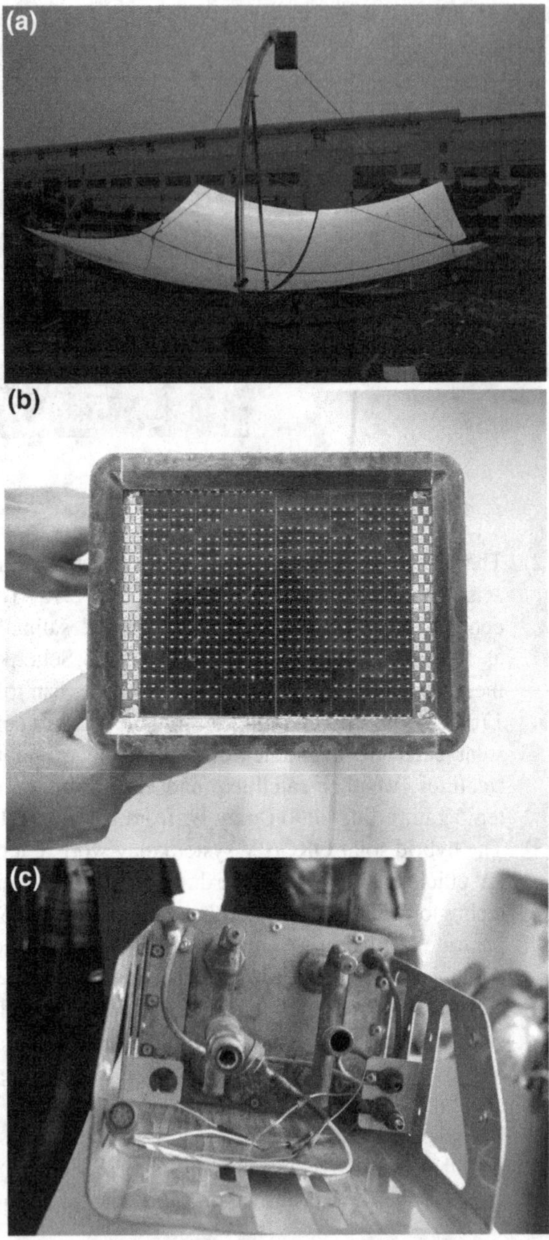

Fig. 2.12 Concentrated PV/T system: **a** overview; **b** front of the PV receiver; **c** back of the PV receiver

2.1.3 Osmosis-Driven Solar ORC

Figure 2.13 shows the scheme of the solar ORC system using semi-permeable membrane for tri-generation applications. The system consists of expander, condenser, evaporator, absorber, semi-permeable membrane and generator. The membrane is only permeable to the solvent. The absorber contains dilute solution, and the generator contains stronger solution. The stronger solution develops a higher osmotic pressure, which makes the solvent move from dilute solution towards it spontaneously.

There are two basic modes. Mode I can be described as the combined cooling and power (CCP) mode. Valves 1, 3 and 5 are open. The solvent is vaporized in the generator. The vapor of high pressure and temperature goes into the expander and exports power. The solvent leaves the expander with lower pressure and temperature, and is cooled by water in the condenser. The solvent from the condenser is throttled, and then flows into the evaporator where it is vaporized and chills the water. The vapor is absorbed by the solution in the absorber. Finally, the solvent experiences osmotic flow through the membrane and goes into the generator. The solvent is heated and circulates in such a way. In this mode, the system provides both cooling and power. The solution temperature and the solvent mass fraction distributions in the generator are shown in Fig. 2.14. 'h', and 'l' denote high value and low value respectively. The arrow denotes the direction from the high value to the low value. Take NH_3/NaSCN solution for example. NH_3 is heated on the way to

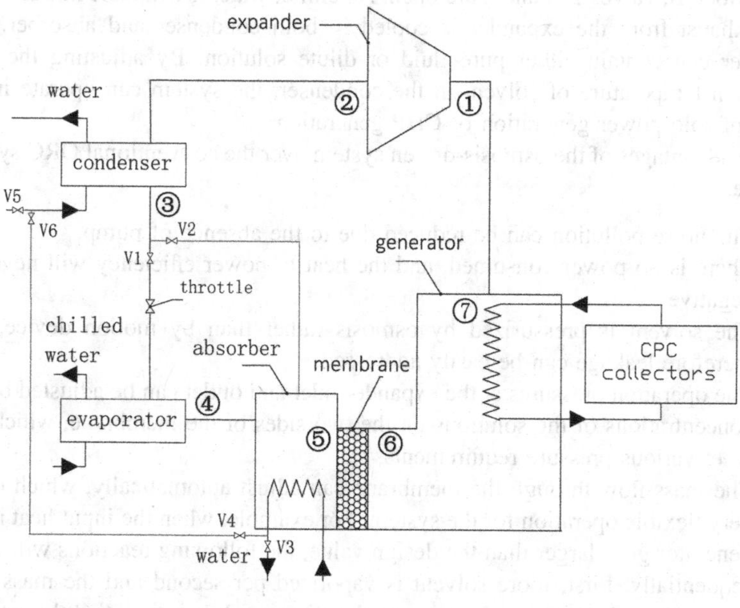

Fig. 2.13 Solar tri-generation system based on semipermeable membrane

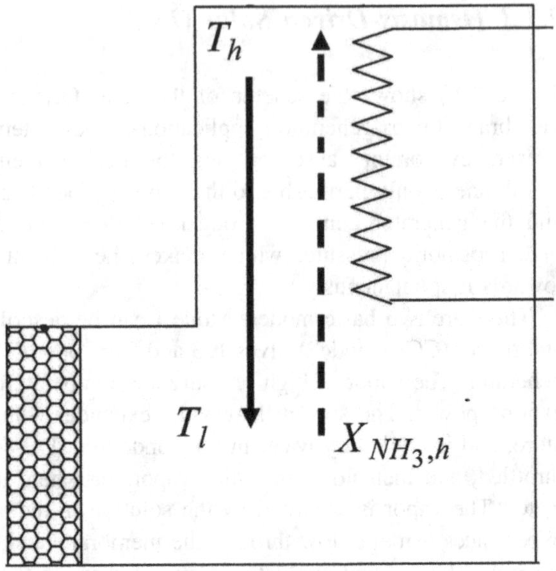

Fig. 2.14 Parameter distribution in the generator

the surface. And the solution density decreases significantly as the temperature increases, which will be validated in the following section. Therefore, the solution has a much higher temperature at the surface than that near the membrane. In both absorber and generator, NH_3 and NaSCN are transferred in a convective diffusion manner.

In Mode II, valves 2, 4 and 6 are open. No chiller water is available in this mode. The exhaust from the expander is cooled in both condenser and absorber. The absorber can contain either pure fluid or dilute solution. By adjusting the condensation temperature of solvent in the condenser, the system can operate in the mode of solo power generation or CHP generation.

The advantages of the osmosis-driven system over the conventional ORC system include:

(1) The noise pollution can be reduced due to the absence of pump.
(2) There is no power consumed, and the heat to power efficiency will never be negative.
(3) The solvent is pressurized by osmosis rather than by motion device, and therefore leakage can be easily addressed.
(4) The operation pressures at the expander inlet and outlet can be adjusted by the concentrations of the solutions on the two sides of the membrane, which can meet various pressure requirements.
(5) The mass flow through the membrane can adjust automatically, which offers very flexible operation for the system. For example, when the input heat in the generator goes larger than the design value, the following reactions will occur sequentially. First, more solvent is vaporized per second and the mass flow rate of the solvent through the expander becomes larger than that through the

membrane. Second, the solution concentration in the boiler is increased and that in the absorber is decreased by the unequal mass flow rates of the solvent. Third, the mass flow rate through the membrane is increased by the larger concentration difference between the two side solutions. Finally, the mass flow rates through the expander and the membrane are equal and the system reaches steady state again. More power is output automatically as the input heat increases.

The potential of the semi-permeable membrane for applying in solar ORC systems can be analyzed as follows:

(1) Semi-permeable membranes have been successfully applied in nanofiltration (NF), reverse osmosis (RO), and forward osmosis (FO) technologies. The physical process for semi-permeable membranes is osmosis. One example for this is water movement from the root of a tree to its top as simply illustrated by Fig. 2.15, in which the upper side pressure of the membrane is higher. The membrane's permeability is determined by the size of the pores, which shall be large enough to let small particles pass freely but small enough to inhibit the passage of larger molecules. NF membranes have nominal pore sizes around 1 nm. They can remove large organic molecules and a range of salts. RO/FO membranes have smaller pore sizes than NF membranes. They can remove many types of molecules and ions from solutions. RO and FO are among the most important technologies for water desalination and purification nowadays. And the membrane technology has been commercialized.
(2) Traditional NF/RO/FO membranes are applied for systems in which the solvent is water. However, in the past 10 years the possibility of using NF/RO/FO

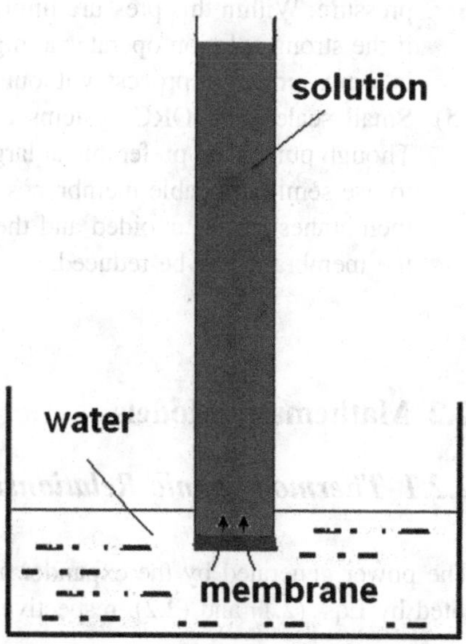

Fig. 2.15 Pressurization due to osmosis

membranes for non-aqueous solvent systems has increased significantly [32]. It is now possible to use the membranes to treat organic fluids. Applications include diafiltration of acetone-methanol, nanofiltration of benzene-lactam and removal of heavies from organic solvents, etc. [33, 34].

(3) The proposed system favorably uses organic solvents. The optional working fluids can be NH_3/NaSCN, NH_3/$LiNO_3$, R22/DMETEG (Dimethyl Ether of Tetraethylene Glycol), R22/DMF(Dimethyl Formamide), R22/Mesitylene, R21/DMETEG, R21/DMF, R134a/DMF, etc. These fluids are used in absorption refrigeration systems. For each pair of working fluids, the solvent is of low boiling point and small molecular diameter, and the solute is of higher boiling point and larger molecular diameter. By selecting a suitable pore size, it is expected that the membrane will be permeable to the solvent and impermeable to the solute. Particularly, NH_3 is similar to H_2O in many ways. The diameters of the two molecules are close (about 0.4 nm). Both NH_3 and H_2O exhibit hydrogen bonding. Semi-permeable membranes in water purification may be directly used in the solar ORC system.

(4) Osmotic pressure builds up due to the concentration difference between the solutions on the two sides of the membrane. The osmotic pressure is not the real pressure of a solution, but the minimum pressure required to nullify osmosis. In a conventional RO process, the operation pressure of the solution must be higher than the osmotic pressure to prevent the inward flow of water across the membrane. In the proposed system, there is no pressure higher than the osmotic pressure. And the membrane undergoes the FO process. The osmotic pressure makes the solvent move from a dilute solution to a strong solution of high solute concentration, in the direction that tends to equalize the solute concentrations on the two sides. In theory, this movement is spontaneous as long as the strong solution operates at pressure less than the osmosis pressure. Within this pressure limit, the solvent can move in such a way even if the strong solution operates at higher pressure than the dilute solution. This is a pressurization process without pump.

(5) Small scale solar ORC systems are characterized by low mass flow rates. Though pumps are preferable at large mass flow rates, it may be more efficient to use semi-permeable membranes at small rates, for which large amounts of membranes can be avoided and the irreversibility of the mass transfer across the membrane can be reduced.

2.2 Mathematic Models

2.2.1 Thermodynamic Relationships in the ORC

The power generated by the expander and that consumed by the pump are calculated by Eqs. (2.1) and (2.2), respectively,

$$w_t = m(h_{t,i} - h_{t,o}) \tag{2.1}$$

$$w_p = m(h_{p,o} - h_{p,i}) \tag{2.2}$$

The isentropic efficiency for the expander and the pump is defined by Eqs. (2.3) and (2.4),

$$\varepsilon_t = \frac{h_{t,i} - h_{t,o}}{h_{t,i} - h_{t,os}} \tag{2.3}$$

$$\varepsilon_p = \frac{h_{p,os} - h_{p,i}}{h_{p,o} - h_{p,i}} \tag{2.4}$$

where *os* represents the ideal thermodynamic process.

When a regenerator is settled, the vapor leaving the expander transfers heat to the liquid from the pump. The regenerator is efficiency is expressed as

$$\varepsilon_r = \frac{h_{r,lo} - h_{r,li}}{h_{r,vi} - h_{r,vo(T_{r,vo}=T_{r,li})}} \tag{2.5}$$

The enthalpy $h_{v,o(T_{v,o}=T_{l,i})}$ is obtained by assuming that the vapor outlet temperature is equal to the liquid inlet temperature. It is an assumptive value but not the real enthalpy.

The energy required in the heating process of the ORC is calculated by the enthalpy increment of the organic fluid from the regenerator to the expander.

$$q = m(h_{t,i} - h_{r,lo}) \tag{2.6}$$

The ORC efficiency is defined by the ratio of the net power output to the heat supplied,

$$\eta_{ORC} = \frac{w_t \cdot \varepsilon_g - w_p}{q} \tag{2.7}$$

For the solar ORC using semipermeable membrane and solution, the energy required in the heating process is

$$q = m[h_{t,i} - (\bar{h}_a/M)] \tag{2.8}$$

where $\bar{h}_a$ is the partial molar enthalpy of the solvent in the solution, M is molecular weight of the solvent.

When the solution is NH_3/NaSCN, the saturation pressure of NH_3 can be calculated by [34]

$$\ln p = 15.7266 - 0.298628X - [2548.65 + 2621.92(1 - X)^3]/T \quad (2.9)$$

where X is the mass fraction of NH_3. The units for p and T are kPa and K.

The specific enthalpy of the solution can be expressed as

$$h(T, X) = H_1(X) + H_2(X)(T - 273.15) + H_3(X)(T - 273.15)^2 + H_4(X)(T - 273.15)^3 \quad (2.10)$$

where

$$H_1(X) = 79.72 - 1072X + 1287.9X^2 - 295.67X^3 \quad (2.11)$$

$$H_2(X) = 2.4081 - 2.2814X + 7.9291X^2 - 3.5137X^3 \quad (2.12)$$

$$H_3(X) = 0.01255X - 0.04X^2 + 0.0306X^3 \quad (2.13)$$

$$H_4(X) = [-0.0333X + 0.1X^2 - 0.0333X^3]10^{-3} \quad (2.14)$$

The partial molar enthalpy of the solvent in the solution is then obtained by

$$\bar{h} = M\partial h(T, X)/\partial X \quad (2.15)$$

The density of the solution can be calculated by

$$\rho(T, X) = A + B(T - 273.15) + C(T - 273.15)^2 \quad (2.16)$$

where

$$A = 1707.519 - 2400.4348X + 2256.5083X^2 - 930.0637X^3 \quad (2.17)$$

$$B = -3.6341X + 5.4552X^2 - 3.1674X^3 \quad (2.18)$$

$$C = 10^{-3}(5.1X - 3.6X^2 - 5.4X^3) \quad (2.19)$$

The osmotic pressure Π of the solution fulfills the following relationship

$$-RT \ln \gamma_s x_s = \int_p^{p+\Pi} v dp \quad (2.20)$$

where R is the gas constant; γ_s is the activity coefficient, often assumed to be 1.0; x_s is the molar fraction of the solvent; p is the saturation pressure of pure solvent at T, v is the molar volume. This equation can be deduced when considering equilibrium between the solution and pure solvent. For most liquids, v varies very slightly with p. For example, the volume of NH_3 at temperature of 300 K and pressure of 1.2 MPa is 28.32 mL/mol, and at 300 K and 2.4 MPa it is 28.27 mL/mol. Therefore incompressible solution can be assumed, and

$$\Pi = -(RT/v)\ln(\gamma_s x_s) \approx -(RT/v)\ln x_s \tag{2.21}$$

When determining the mole fraction of solvent, it is necessary to include the ionization of salts. For example one mole of NaSCN ionizes to two moles of ions, and the mole fraction of ammonia reduces accordingly.

2.2.2 *Efficiency in Solar Energy Collection*

2.2.2.1 Solar System Using Commercial Collectors

For the solar ORCs with CDVG and semipermeable membrane, commercial solar collectors are adopted. The thermal efficiency of a commercial FPC or CPC collector is generally expressed by

$$\eta_c(T) = \eta_{c,0} - \frac{A}{G}(T - T_a) - \frac{B}{G}(T - T_a)^2 \tag{2.22}$$

The solar collector modules available on the market have effective area between 1.0 and 2.0 m^2. Their thermal efficiency can be calculated by Eq. (2.22). In a solar ORC system tens or hundreds square meters of collectors are usually required, the temperature difference between neighboring collectors is supposed to be small. To calculate the overall collector efficiency it is reasonable to make the assumption that the average operating temperature of the collector changes continuously from one module to another.

The working fluid in the collector is mostly at liquid-phase or binary phase. For the binary phase region, the temperature is constant and it is easy to calculate the collector efficiency. For the liquid phase region, in order to reach an outlet temperature $T_{f,o}$ of the fluid with an inlet temperature $T_{f,i}$, the required collector area is obtained by

$$S_l = \int_{T_{f,i}}^{T_{f,o}} \frac{m_f C_{p,f}(T)}{\eta_c(T) G} dT \tag{2.23}$$

The heat capacity of a liquid can be expressed by a first order approximation

$$C_p(T) = C_{p,0} + \alpha(T - T_0) \tag{2.24}$$

With $c_1 = A/G$, $c_2 = B/G$, the collector area according to Eqs. (2.22), (2.23) and (2.24) is calculated by

$$S_l = \frac{m_f}{c_2 G(\theta_2 - \theta_1)}\left[(C_{p,a} + \alpha\theta_1)\ln\frac{(T_{f,o} - T_a - \theta_1)}{T_{f,i} - T_a - \theta_1} + (C_{p,a} + \alpha\theta_2)\ln\frac{\theta_2 - T_{f,i} + T_a}{\theta_2 - T_{f,o} + T_a}\right] \tag{2.25}$$

θ_1 and θ_2 are the arithmetical solutions of Eq. (2.26) ($\theta_1 < 0,\ \theta_2 > 0$).

$$\eta_o - c_1\theta - c_2\theta^2 = 0. \tag{2.26}$$

$$C_{p,a} = C_{p,0} + \alpha(T_a - T_0) \tag{2.27}$$

The thermal efficiency of the collectors with liquid is calculated by

$$\eta_{c,l} = \frac{m_f(h_{l,o} - h_{l,i})}{GS_l} \tag{2.28}$$

The thermal efficiency of the collectors with working fluid in the binary-phase and the overall collector system are calculated by Eqs. (2.29) and (2.30), respectively,

$$\eta_{c,v} = \frac{m_f(h_{b,o} - h_{b,i})}{GS_b} \tag{2.29}$$

$$\eta_c = \frac{m_f(h_{b,o} - h_{l,i})}{G(S_l + S_b)} \tag{2.30}$$

2.2.2.2 Solar ORC with PV Module

For the solar ORC with PV module, the mathematic models are more complicated because no commercial PV-CPC collector is available at present. In this case, Eq. (2.22) is not applicable.

The total radiative heat loss from the PV-absorber can be written in terms of A_S, T_S and T_L [35]:

$$q_{rad,S\to R+L} = \varepsilon_{eff} A_S \sigma (T_S^4 - T_L^4) \tag{2.31}$$

$$\varepsilon_{eff} = \varepsilon_{eff\ SL} + \varepsilon_{eff\ SL}\varepsilon_{eff\ RL} / (\varepsilon_{eff\ SL} + \varepsilon_{eff\ RL}) \quad (2.32)$$

where $\sigma = 5.67 \times 10^{-8}$ W/(m^{2}K) is the Stefan-Boltzmann constant; A_S is the absorber area; ε_{eff} is the effective emissivity; The subscripts of *S, L, R* represent the absorber, glass cover and reflector, respectively. *SL* and *RL* represent the corresponding radiative heat transfer between surfaces *S* and *L*, and surfaces *R* and *L*. The emissivities of PV cells and the absorber are assumed to be uniform.

The overall convective heat transfer by the air inside the CPC is given by [36]

$$q_{conv,SL} = A_S U_S (T_S - T_L)/(1 + C^{-0.6}) \quad (2.33)$$

$$U_S = [(g\beta/v^2) l^3 \Delta T_S \Pr]^a \cdot c \cdot (k/l) \quad (2.34)$$

where U_S is heat transfer coefficient for the absorber surface; k is the conductivity; l is the length of the absorber; *Pr* is the Prandtl number; a and c are constants for which the values have been suggested by [37]; g is earth's acceleration; β is the volume coefficient of expansion; v is the kinematic viscosity; C is the concentration ratio. Heat flow rate from L to the ambient is calculated by

$$q_{L-a} = \varepsilon_L A_L \sigma (T_L^4 - T_{sky}^4) + A_L U_a (T_L - T_{air}) \quad (2.35)$$

Figure 2.16 shows thermal energy network for a single PV-CPC module. The total irradiation absorbed by a single PV-CPC module is

$$q_{S,sol} = G A_L \gamma \tau_L \rho_{R,sol} \alpha_{S,sol} \quad (2.36)$$

where γ is fraction of total irradiation accepted by the CPC; τ_L is solar transmissivity of cover; $\rho_{R,sol}$ is solar reflectivity of reflector; $\alpha_{S,sol}$ is solar absorptivity of PV cells and absorber. And various energy fluxes are given by

$$\begin{aligned} q_{L-a} &= h_{L-a} A_L (T_L - T_a) \\ &= \varepsilon_L A_L \sigma (T_L^4 - T_{sky}^4) + A_L U_a (T_L - T_{air}) \end{aligned} \quad (2.37)$$

$$\begin{aligned} q_{S-L} &= h_{r,S-L} A_S (T_S - T_L) + h_{c,S-L} A_S (T_S - T_L) \\ &= \varepsilon_{eff} A_S \sigma (T_S^4 - T_L^4) + A_S U_S \Delta T_S \end{aligned} \quad (2.38)$$

$$q_{S-f} = h_{S-f} A_f (T_S - T_f) \quad (2.39)$$

$$q_{f-a} = U_{f-a} A_f (T_f - T_a) \quad (2.40)$$

A_f is the heat transfer area of tube; U_{f-a} is the coefficient for heat loss through insulation material; $h_{r,S-L}$ is the coefficient of radiative heat transfer while $h_{c,S-L}$ is the coefficient of convective heat transfer.

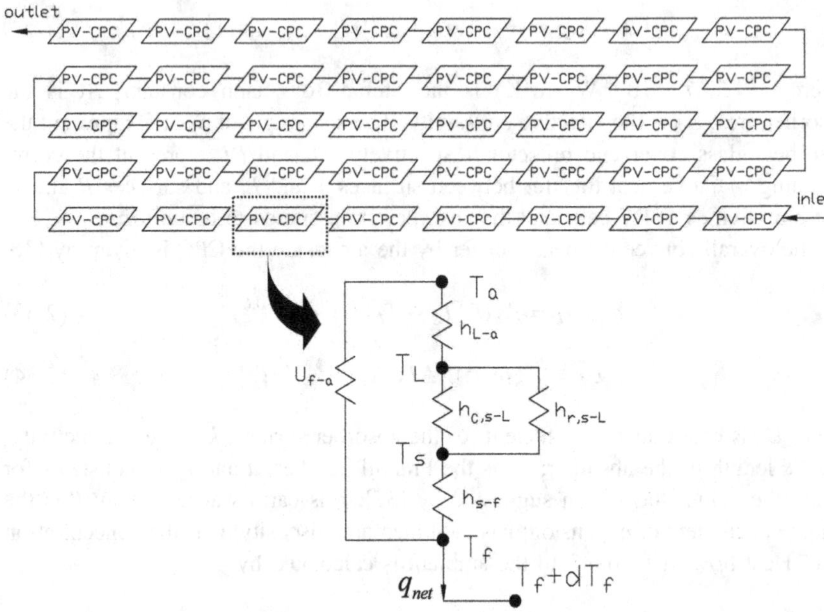

Fig. 2.16 Thermal energy network for single PV-CPC module and connection of PV-CPC modules in series. Reprinted from Ref. [18], Copyright 2010, with permission from Oxford University Press

The energy balances among the cover, absorber, PV cell, and working fluid can be expressed by

$$q_{L-a} - q_{S-L} = 0 \tag{2.41}$$

$$q_{S,sol} - q_{S-L} - q_{S-f} - p_{cell} = 0 \tag{2.42}$$

$$q_{S-f} - q_{f-a} - q_{net} = 0 \tag{2.43}$$

Heat transfer from the absorber to the working fluid is calculated by

$$q_{S-f} = U_{tube}\pi D(T_S - T_f) \tag{2.44}$$

where U_{tube} is the heat transfer coefficient between the wall and working fluid; D is the diameter of the tube. U_{tube} is determined according to Ref. [35].

Electricity produced by PV cells is obtained by

$$p_{cell} = GA_L \cdot \eta_{PV}(T) \cdot x \tag{2.45}$$

where x is cover ratio of PV cells.

$$\eta_{PV}(T) = \eta_{pv,0} - c(T - 25)\eta_{pv,0} \tag{2.46}$$

where $\eta_{pv,0}$ is the maximum PV efficiency on condition of AM1.5, 1,000 W/m^{-2}, 25 °C; c is the temperature coefficient of maximum power produced. The collector efficiency regarding net heat available for the working fluid is

$$\eta_c(T) = q_{net}/(GA_L) \tag{2.47}$$

PV-CPC modules in series are shown in Fig. 2.16. The differential equations for PV-CPC collector area and working fluid enthalpy increment are expressed by

$$q_{net} = \eta_c(T_S)GdA_L = m_f C_{p,f}(T_f)dT_f \tag{2.48}$$

$$\eta_c(T_S)GdA_L = m_f(h_{f,v} - h_{f,l})dX \tag{2.49}$$

Equation (2.48) is for liquid region and Eq. (2.49) for binary region of working fluid. $H_{f,v}$ or $H_{f,l}$ is the enthalpy of the saturated vapor or saturated liquid; X is the dryness.

2.2.3 Solar ORC System Efficiency

The overall electricity efficiency of the solar ORC is expressed by

$$\eta_{sys} = \eta_{ORC} \cdot \eta_c \tag{2.50}$$

2.3 Results and Discussion

2.3.1 Performance of Solar ORC with CDVG

Figure 2.17 shows the variations of overall collector efficiency with the evaporation temperature when using SSE and CDVG, respectively. The environment temperature is 20 °C, which keeps constant in the following analysis. The condensation temperature of the ORC is 25 °C. The organic fluid is R123, which is heated by the collectors from subcooled state to saturated vapor state. CPC collectors are adopted. The collector efficiency is proposed according to some product information [38]. The first heat loss coefficient A is 0.82 W m^{-2} °C^{-1}, the second heat loss coefficient B is 0.0064 W m^{-2} °C^{-2}, and the maximum efficiency is 0.661. The incident insolation is

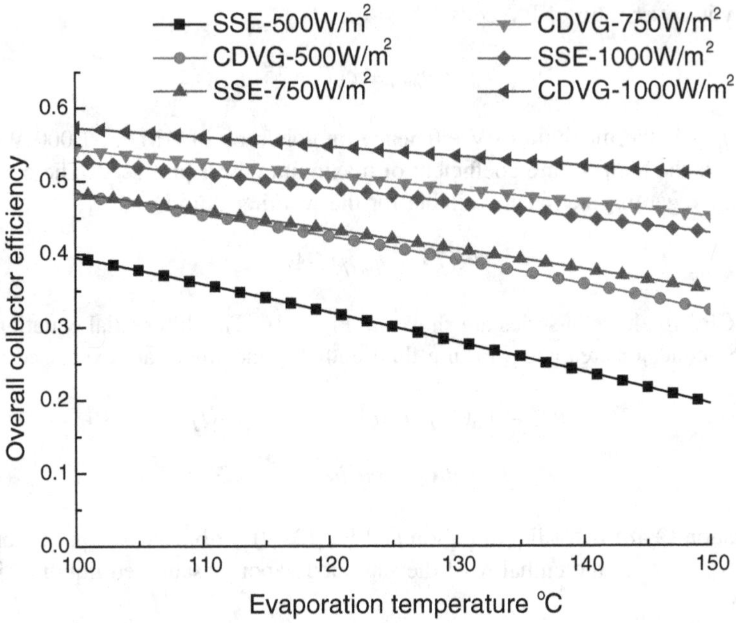

Fig. 2.17 Overall collector efficiency varying with the evaporation temperature at a condensation temperature of 25 °C

500, 750 and 1,000 W m^{-2}. For the solar ORC with SSE, counter-current concentric tubes are used as the heat exchanger (evaporator). The inner and outer diameters of the tubes are 25 and 45 mm. The mass flow rates of R123 and conduction oil through one tube are 0.1 and 0.9 kg/s, with a total heat exchanger area of 62.8 m^2. Based on these conditions, the inlet and outlet temperatures of the oil are 118.0 and 108.1 °C when the evaporation temperature of R123 is 100 °C. Under this evaporation temperature, the overall collector efficiencies of the solar ORC with SSE and CDVG are 0.528 and 0.573 for irradiation of 1,000 W m^{-2}. The relative increment of the latter by the former is about 8.5 %. The efficiencies are 0.484 and 0.543, 0.396 and 0.482 corresponding to irradiation of 750 and 500 W m^{-2}. As the irradiation gets weaker, the relative increment in efficiency with CDVG becomes larger. Meanwhile, when the evaporation temperature is 150 °C, the relative increment is 65.0, 28.8 and 18.7 % for irradiation of 500, 750 and 1,000 W m^{-2}. The relative increment is enlarged at higher evaporation temperature. The advantages of solar ORC with CDVG over the conventional SSE based system are significant in regard to the heat collection efficiency.

Figure 2.18 shows the variations of overall collector efficiency with the evaporation temperature when the condensation temperature is 55 °C. Compared with 25 °C, a higher condensation temperature results in a lower overall collector efficiency for the CDVG based solar ORC. The efficiency at the evaporation temperature of 100 °C falls down to 0.471, 0.535 and 0.566 for irradiation of 500, 750 and

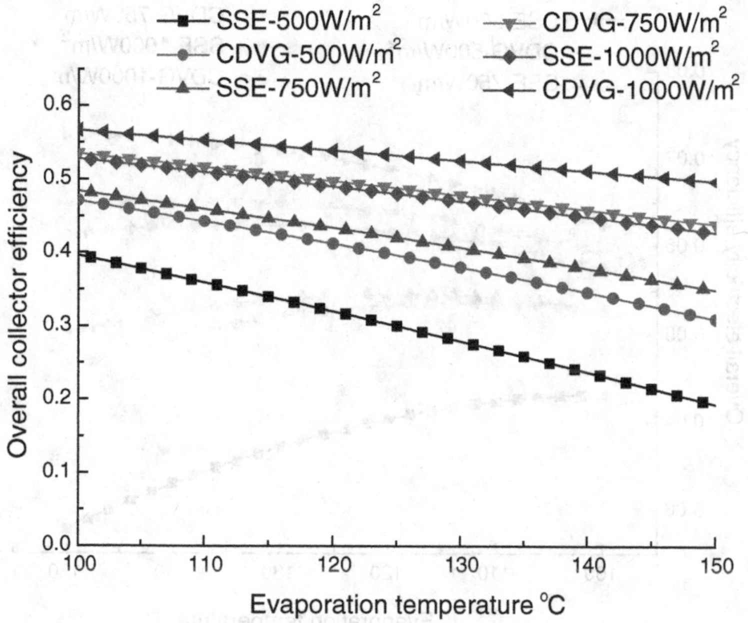

Fig. 2.18 Overall collector efficiency varying with the evaporation temperature at a condensation temperature of 55 °C

1,000 W m^{-2}, which is attributed to a higher average operation temperature of the collectors. On the other hand, the efficiency for the solar ORC with SSE is almost unchanged with the condensation temperature. Given a high mass flow rate, the inlet and outlet temperatures of the conduction oil vary slightly with R123 condensation temperature. For example, they are 117.3 and 108.8 °C at the evaporation temperature of 100 °C and condensation temperature of 55 °C. The results indicate that the superiority of solar ORC with CDVG turns weaker at higher condensation temperature.

Figures 2.19 and 2.20 depict the variations of the solar thermal electricity efficiency with the evaporation temperature. The efficiencies of the expander, generator, pump and regenerator are assumed to be 0.75, 0.85, 0.50 and 0.70, respectively. The curves show that the system electricity efficiency first climbs and then drops down with increment in the evaporation temperature. The heat to power efficiency of the ORC is improved by larger temperature difference between the hot and the cold sides, according to the second law of thermodynamics. But the collector efficiency decreases with increment in the operation temperature. Due to this tradeoff there shall be an optimal evaporation temperature ($T_{eva,opt}$) at which the overall electricity efficiency reaches the maximum ($\eta_{sys,\max}$). Table 2.3 lists the $T_{eva,opt}$ and $\eta_{sys,\max}$ on different conditions of irradiation and condensation temperature. $T_{eva,opt}$ increases with the increment in solar irradiation and condensation temperature. $T_{eva,opt}$ for the solar ORC

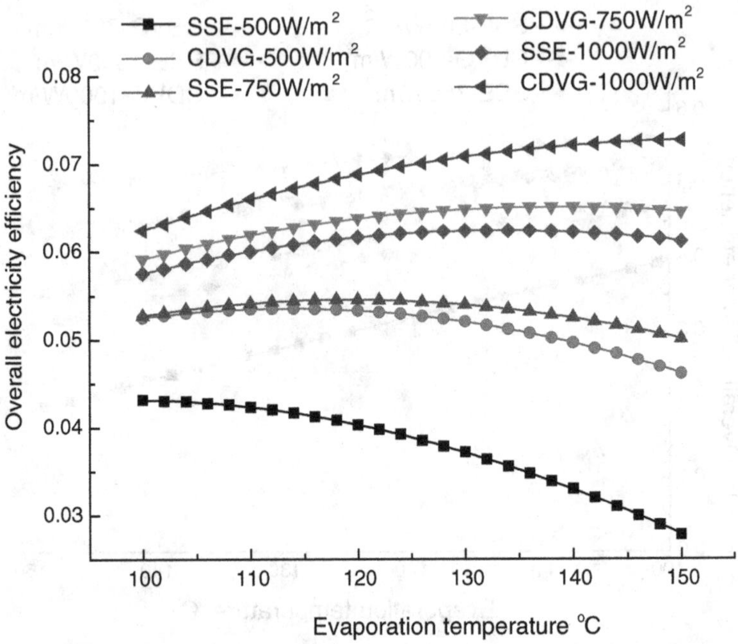

Fig. 2.19 Overall electricity efficiency of the solar ORC varying with the evaporation temperature at a condensation temperature of 25 °C

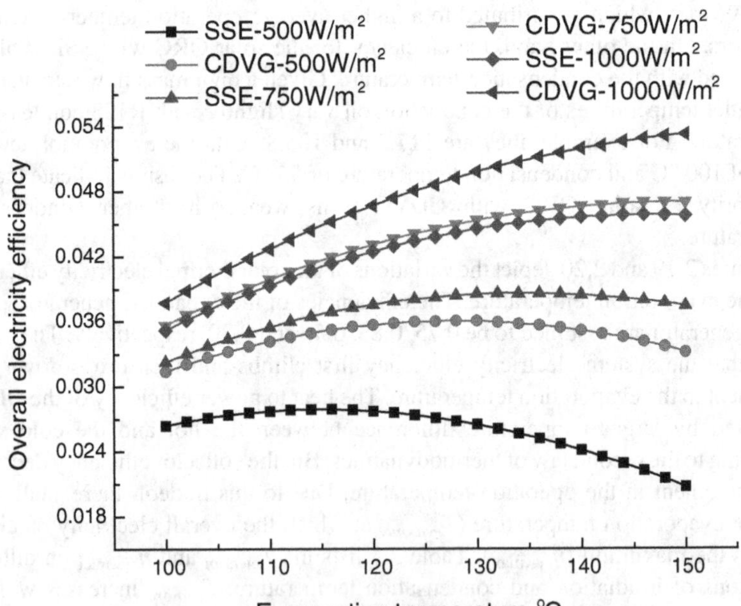

Fig. 2.20 Overall electricity efficiency of the solar ORC varying with the evaporation temperature at a condensation temperature of 55 °C

Table 2.3 Optimal evaporation temperature and corresponding efficiency of the solar ORC on different conditions

Condensation temperature (°C)	Irradiation (W m^{-2})	Solar ORC with SSE		Solar ORC with CDVG	
		$T_{eva,opt}$ (°C)	$\eta_{sys,opt}$ (%)	$T_{eva,opt}$ (°C)	$\eta_{sys,opt}$ (%)
25	500	<100	>4.31	113	5.36
	750	118	5.46	136	6.51
	1,000	130	6.24	149	7.27
55	500	115	2.79	128	3.60
	750	133	3.86	149	4.70
	1,000	144	4.58	>150	>5.33

with CDVG is higher than that with SSE on the same conditions. $\eta_{sys,\max}$ is also higher, and the relative increment is about 16.3–29.0 %.

Only the CPC collector is considered in Figs. 2.17, 2.18, 2.19 and 2.20. The FPC collector seems more cost-effective than the CPC as mentioned in Sect. 1.1. The comparison of thermal efficiencies of some commercial FPC and CPC collectors is shown in Fig. 2.21. For the FPC collector, the first heat loss coefficient A is 3.157 W m^{-2} °C^{-1}; the second heat loss coefficient B is 0.014 W m^{-2} °C^{-2}, and the maximum efficiency is 0.857. The dash dot lines mean the intersection of the efficiencies of the FPC and CPC. The intersection temperature is 58, 74 and 89 °C correspondingly for irradiation of 500, 750 and 1,000 W m^{-2}. The FPC can offer a higher

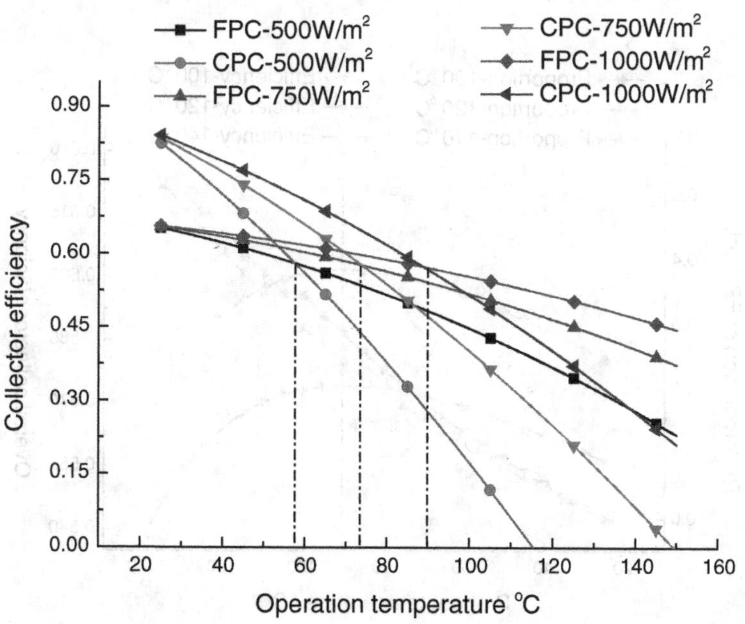

Fig. 2.21 Variations of the FPC and CPC efficiencies with operation temperature

collector efficiency at temperature below the intersection. So the overall collector efficiency will be improved if the organic fluid is preheated properly by FPCs.

The variations of the required FPC proportion and the overall collector efficiency with the preheat temperature is displayed in Fig. 2.22. The fluid is preheated by FPCs and is further heated by CPCs. The proportion is the ratio of the FPC area and the total area of the collector system. The irradiation is 1,000 W m^{-2}. The condensation temperature is 25 °C. The fluid leaving the CPCs is assumed to be at saturated vapor state. The FPC proportion increases with the increment in the preheat temperature, which denotes the temperature at the FPCs outlet. For each efficiency curve, the overall collector efficiency first climbs as the preheat temperature goes up. However, the efficiency does not increase in a monotone way. There exists an optimal preheat temperature, at which the overall collector efficiency reaches the maximum. According to fundamental mathematics, the derivative of overall collector efficiency with respect to the preheat temperature is equal to zero at the optimal point ($T_{FPC,opt}$). $T_{FPC,opt}$ shall fulfill the requirement that the local efficiency at FPCs outlet be equal to that at the CPCs inlet. It means $T_{FPC,opt}$ is equal to the intersection temperature as marked in Fig. 2.21.

A proof by contradiction can be performed. If $T_{FPC,opt}$ should be lower than the intersection temperature, then a tiny area ΔS of the CPC collector could be replaced by the FPC collector, and the overall collector efficiency would be increased. If $T_{FPC,opt}$ should be higher than the intersection temperature, then a tiny area ΔS of the FPC collector could be replaced by the CPC collector, resulting in a higher overall collector efficiency. For both cases the efficiency at $T_{FPC,opt}$ is not maximum. Therefore, $T_{FPC,opt}$ should be equal to the intersection temperature.

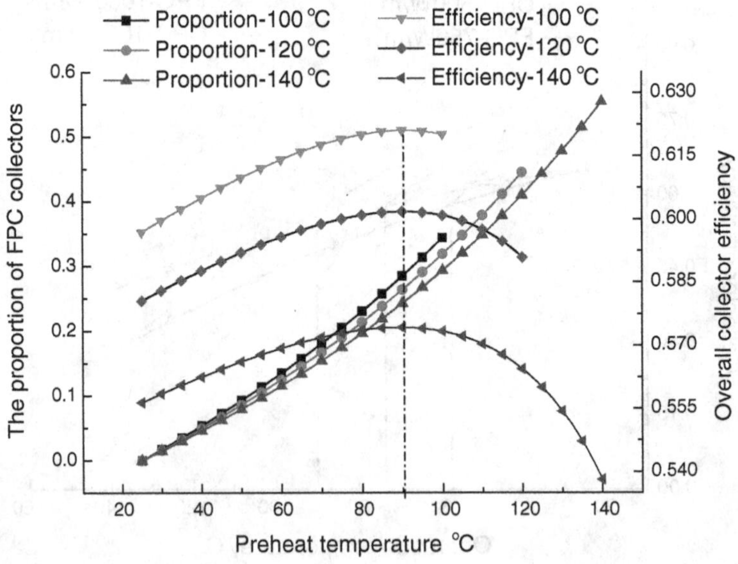

Fig. 2.22 FPC proportion and overall collector efficiency varying with the preheat temperature

The overall collector efficiency at $T_{FPC,opt}$ is about 0.62, 0.60 and 0.57 for the evaporation temperature of 100, 120 and 140 °C respectively. The corresponding FPC proportion at $T_{FPC,opt}$ is about 29.0, 26.8 and 24.7 %. Compared with the solo CPC system (with a FPC proportion of zero), the two-stage collectors provide more efficient solar energy collection. And the relative increment in the efficiency is about 4 %.

The two-stage collectors are advantageous in the simultaneous processes of heat collection and power conversion from the viewpoint of cost and thermodynamic performance, where the temperature of fluid entering the collector system is much lower than the outlet temperature. However, regarding the heat storage the two-stage collectors may be disadvantageous if only one kind of PCM is used. Figure 2.23 shows the variations of overall efficiencies of the single-stage and two-stage collectors with irradiation in the heat storage process. The FPC proportion is 26.8 %. The heat from both FPCs and CPCs is transferred to a PCM of melting point of about 120 °C. Thus the operation temperatures of the FPCs and CPCs get alike, and are assumed to be 120 °C. As illustrated, the two-stage collectors lead to a lower efficiency in the heat storage process.

To solve this problem, two-stage heat storage units are necessary, in which heat storage PCM (1) and PCM (2) are connected to the FPC and CPC collectors, respectively. Because the heat stored by the PCMs will be further used by the ORC, the heat collection from FPCs in the heat storage process is expected to match the

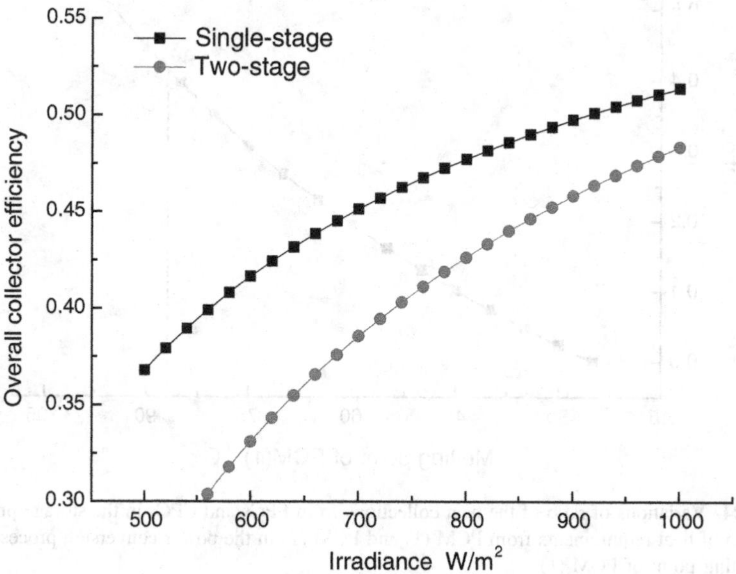

Fig. 2.23 Variations of overall efficiencies of the single-stage and two-stage collectors with irradiation in the heat storage process when only one kind of PCM is used

enthalpy increment of R123 by PCM (1) in the power conversion process. It means the ratio of the heat collections from FPCs and CPCs in the storage process shall be equal to the ratio of heat requirements from PCM (1) and PCM (2) in the power conversion process. Given the FPC proportion, both ratios are correlated to the melting point of PCM (1), as shown in Fig. 2.24. The irradiation is 1,000 W m^{-2}. The evaporation and condensation temperatures are 120 and 25 °C. A higher melting point of PCM (1) results in lower FPC efficiency, and more energy to heat R123 from 25 °C to the melting point. So the ratio of heat collection decreases while the ratio of heat requirement increases with the increment in PCM (1) melting point. The intersection of the curves can offer a balance between the heat collection and heat requirement. When the melting point is lower than the intersection, more heat is stored by PCM (1) than the released, which may cause an inefficient use of the FPCs and heat. And a higher melting point is accompanied by a lower overall collector efficiency in the storage process.

Figure 2.25 shows the variations of overall efficiencies of the single-stage and two-stage collectors with irradiation in the heat storage process when two kinds of

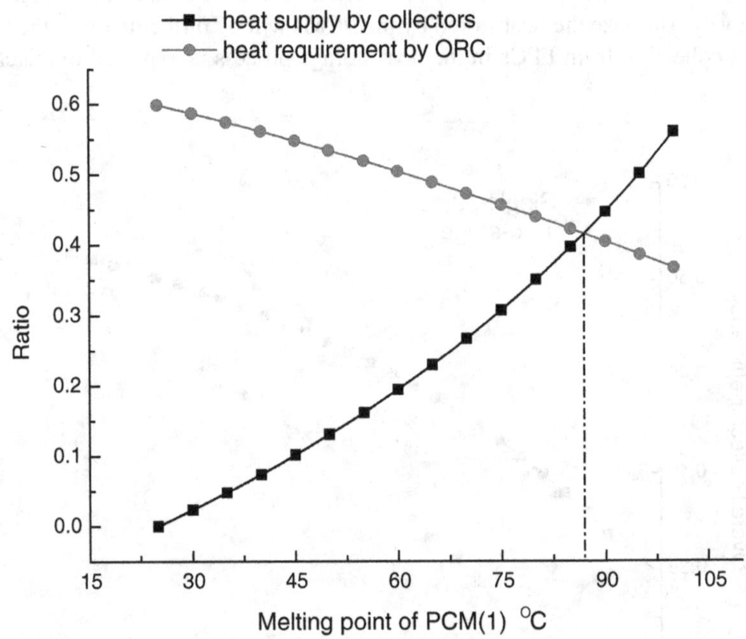

Fig. 2.24 Variations of ratio of the heat collections from FPCs and CPCs in the storage process, and ratio of heat requirements from PCM (1) and PCM (2) in the power conversion process with the melting point of PCM (1)

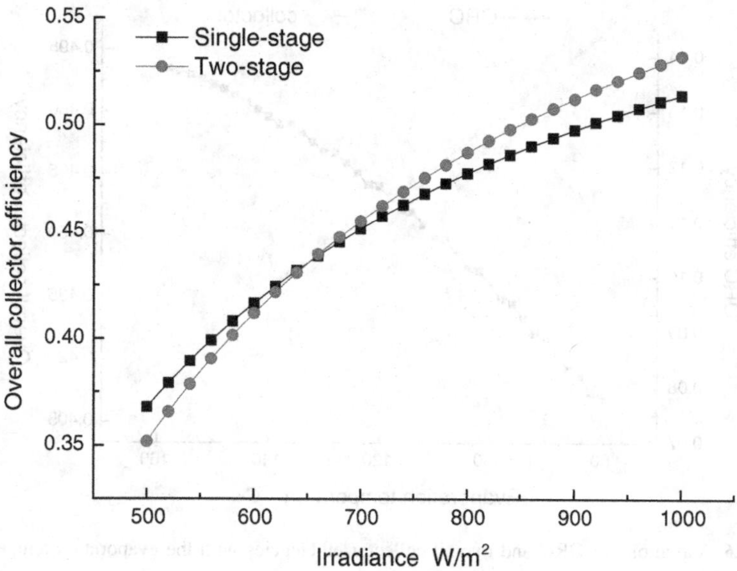

Fig. 2.25 Variations of overall efficiencies of the single-stage and two-stage collectors with irradiation in the heat storage process when two kinds of PCMs are used

PCMs are used. PCM (1) has a melting point of 87 °C. The other conditions are the same to those for Fig. 2.23. The two-stage heat storage units can improve the heat collection efficiency especially when the irradiation is strong.

2.3.2 Performance of Solar ORC with PV Module

Figure 2.26 shows the ORC efficiency and overall collector efficiency varying with R123 evaporation temperature. The PV standard efficiency $\eta_{pv,0}$ is 7.27 %; the PV temperature coefficient c is 0.21 %; the irradiation is 1,000 W m^{-2}. The PV-CPC module has a concentration ratio of 4.9. The absorber has width of 0.05 m, emissivity of 0.1, and absorptivity of 0.9. The efficiencies of the expander, pump, and generator are 0.75, 0.85 and 0.5. The tube has an inner diameter of 0.025 m. The mass flow rate through the tube is 0.2 kg/s. The ORC efficiency increases with the increment in the evaporation temperature. But the overall collector efficiency gets lower when the evaporation temperature rises.

Figure 2.27 shows the average PV efficiency and total system electricity efficiency varying with R123 evaporation temperature. The average PV efficiency turns lower when the evaporation temperature rises. Notably, the slope of the PV efficiency curve is smaller than the defined temperature coefficient of maximum power generated of PV cell, e.g. at the point of evaporation temperature of 100 °C, the tangent slope of the curve is 0.0014 $°C^{-1}$, about two thirds of the temperature coefficient

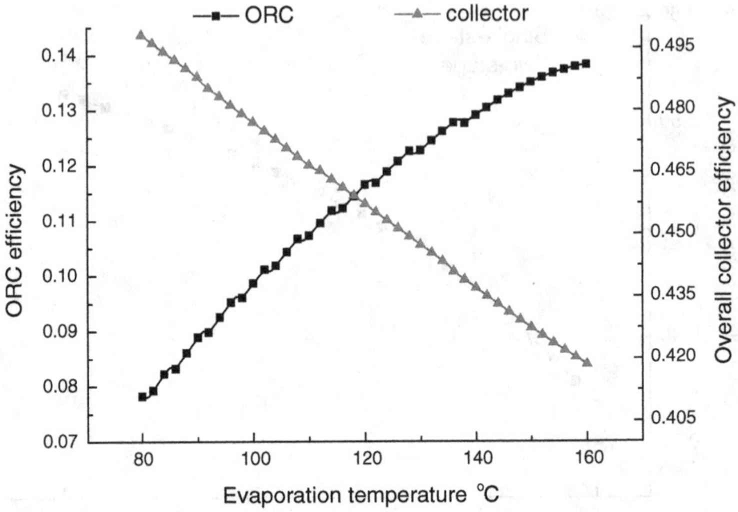

Fig. 2.26 Variations of ORC and overall collector efficiencies with the evaporation temperature

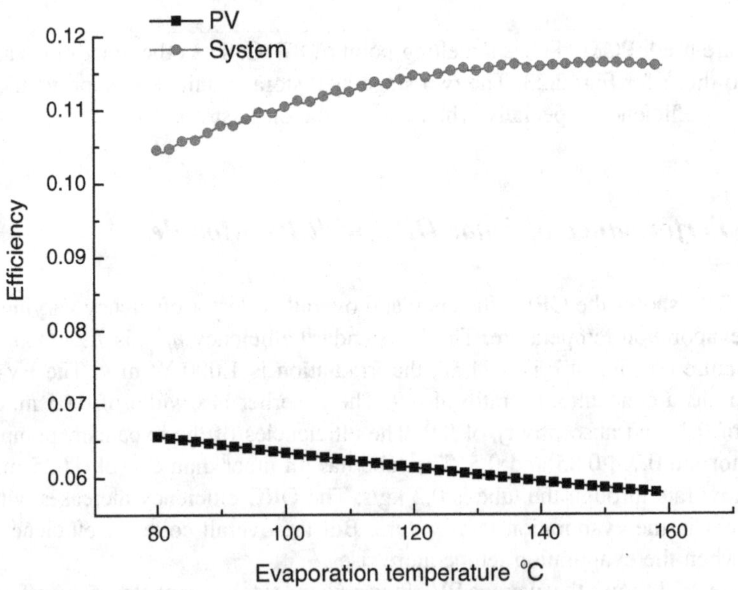

Fig. 2.27 Variations of PV and total system electricity efficiencies with the evaporation temperature

(0.0021 °C^{-1}). This is because that in the PV-CPC system the temperature of one module is different from another and the fluid is heated from condensation temperature to evaporation temperature. The average temperature of the PV-CPC modules is determined by the distribution of lower and higher temperature modules, which will increase more slowly with the increment in the evaporation temperature. The results indicate the influence of the hot side temperature on the efficiency of PV system differs from that on a single cell. The former is slighter than the latter.

In a distinctive manner, the system electricity efficiency first climbs up and then falls down with increment of the evaporation temperature. The increment of evaporation temperature has positive effect on the ORC efficiency, and negative one on the average PV efficiency and overall collector efficiency. On the given conditions, there is a maximum system electricity efficiency. Even if the system efficiency is not at the maximum point, it will be still much higher than the PV efficiency. For example, when the evaporation temperature is 100 °C, the system electricity efficiency is 11.1 %, which is about 53 % higher than the PV efficiency at room temperature (7.27 %).

2.3.3 Performance of Osmosis-Driven Solar ORC

The variation of the osmotic pressure (Π) of NH_3/NaSCN solution with the mass fraction of NH_3 is shown in Fig. 2.28. Π increases steeply when NH_3 mass fraction decreases. It is about 3.9, 8.7 and 14.5 MPa respectively when the fraction is 0.9,

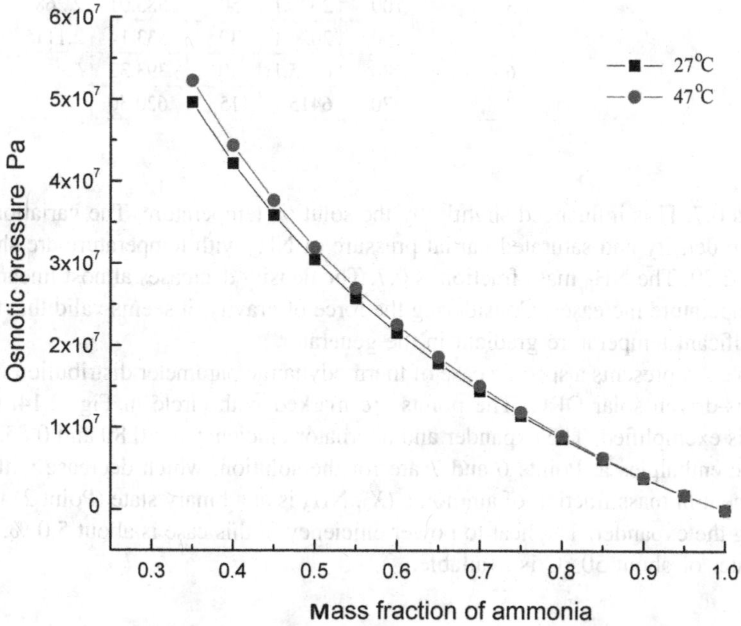

Fig. 2.28 Osmotic pressure variation with the mass fraction of ammonia

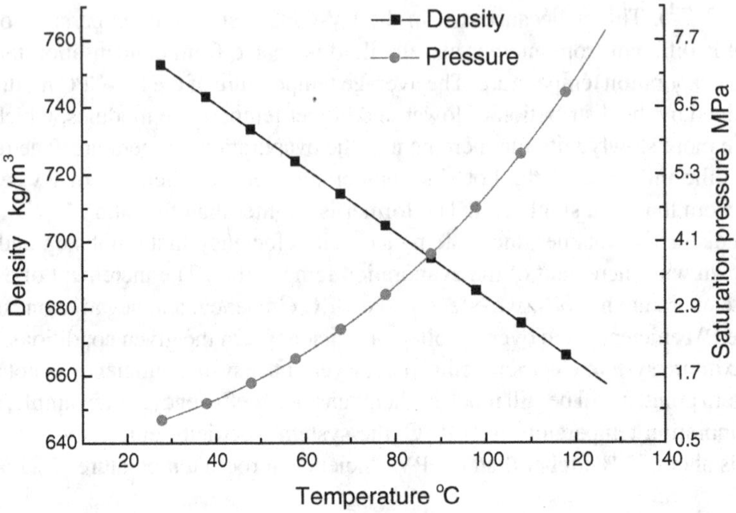

Fig. 2.29 Variations of the solution density and saturated partial pressure of NH_3 with temperature

Table 2.4 A specific parameter distribution in the osmosis-driven solar ORC system

State point	X (%)	*p* (kPa)	*T* (°C)	*h* (kJ/kg)	*s* (kJ/kg/°C)
1	100	6415.1	115	1622.9	5.18
2	100	2026.1	50	1551.2	5.2663
3	100	2026.1	50	583.01	2.2683
5	100	2026.1	40	533.14	2.1115
6	80	6415.1	40	393.32	/
7	70	6415.1	115	620.30	/

0.8 and 0.7. Π is influenced slightly by the solution temperature. The variations of solution density and saturated partial pressure of NH_3 with temperature are shown in Fig. 2.29. The NH_3 mass fraction is 0.7. The density decreases almost linearly as the temperature increases. Considering the force of gravity, it seems valid that there is significant temperature gradient in the generator.

Table 2.4 presents a specific case of thermodynamic parameter distribution in the osmosis-driven solar ORC. The points are marked with circle in Fig. 2.14. CHP mode is exemplified. The expander and alternator efficiency are 0.80 and 0.85. The specific enthalpies at Points 6 and 7 are for the solution, which decrease with the decrement in mass fraction of ammonia (*X*). NH_3 is at a binary state (Point 2) when leaving the expander. The heat to power efficiency in this case is about 5.0 %. And hot water of about 50 °C is available.

References

1. Li J, Pei G, Li Y, Wang D, Ji J (2012) Energetic and exergetic investigation of an organic Rankine cycle at different heat source temperatures. Energy 38:85–95
2. Bernier MA, Bourret B (1999) Pumping energy and variable frequency drives. Am Soc Heat Refrig Air Cond Eng (ASHRAE) J 41(12):37–40
3. Pump technology: improved energy efficiency in refrigeration plants. www.hermetic-pumpen.com/assets/251/FB_ENERGIE_E.pdf. Accessed 31 Dec 2012
4. Liquid refrigerant pumping. www.ashrae.org/File%20Library/docLib/Journal%20Documents/2011%20August/036-043_jekel.pdf. Accessed 31 Dec 2012
5. Lin C (2008) Feasibility of using power steering pumps in small-scale solar thermal electric power systems. Thesis, Massachusetts Institute of Technology
6. Nesbitt B (2006) Handbook of pumps and pumping. Chapter 10: Drivers for pumps. Elsevier, Amsterdam, pp 261–278
7. Electrical motor efficiency. www.engineeringtoolbox.com/electrical-motor-efficiency-d_655.html. Accessed 1 Jan 2013
8. Li J, Pei G, Li Y, Wang D, Ji J (2013) Examination of the expander leaving loss in variable organic Rankine cycle operation. Energy Convers Manage 65:66–74
9. Hu H, Cheng W (2006) Concise thermal physics. Press of University of Science and Technology of China, Hefei, pp 196–197
10. Tahir MBM, Yamada N (2009) Characteristics of small ORC system for low temperature waste heat recovery. J Environ Eng 4:375–385
11. Yamada N, Minami T, Mohamad M (2011) Fundamental experiment of pumpless Rankine-type cycle for low-temperature heat recovery. Energy 36:1010–1017
12. Li J, Pei G, Li Y, Ji J (2013) Analysis of a novel gravity driven organic Rankine cycle for small-scale cogeneration applications. Appl Energy 108:34–44
13. Liu Z, Tao G, Lu L, Wang Q (2014) A novel all-glass evacuated tubular solar steam generator with simplified CPC. Energy Convers Manage 86:175–185
14. Pei G, Li G, Zhou X, Ji J, Su Y (2012) Experimental study and exergetic analysis of a CPC-type solar water heater system using higher-temperature circulation in winter. Sol Energy 86:1280–1286
15. Li X, Dai YJ, Li Y, Wang RZ (2013) Comparative study on two novel intermediate temperature CPC solar collectors with the U-shape evacuated tubular absorber. Sol Energy 93:220–234
16. Ma C (2014) The core competence of solar thermal power technology. In: China CSP investment and financing summit and CSPPLAZA 2014 annual conference, Beijing, 23–24 Aug 2014
17. www.chemcp.com/news/201409/482545.asp. Accessed 25 June 2014
18. Li J, Pei G, Li Y, Ji J (2010) Novel design and simulation of a hybrid solar electricity system with organic Rankine cycle and PV cells. Int J Low Carbon Technol 5:223–230
19. Chow TT (2010) A review on photovoltaic/thermal hybrid solar technology. Appl Energy 87:365–379
20. Ji J, Lu JP, Chow T, He W, Pei G (2007) A sensitivity study of a hybrid photovoltaic/thermal water-heating system with natural circulation. Appl Energy 84:222–237
21. He W, Chow TT, Ji J, Lu J, Pei G, Chan L (2006) Hybrid photovoltaic and thermal solar-collector designed for natural circulation of water. Appl Energy 83:199–210
22. Ji J, Pei G, Chow T, Liu K, He H, Lu J, Han C (2008) Experimental study of photovoltaic solar assisted heat pump system. Sol Energy 82:43–52
23. Mittelman G, Kribus A, Dayan A (2007) Solar cooling with concentrating photovoltaic/thermal (CPVT) systems. Energy Convers Manage 48:2481–2490
24. Garcia-Heller V, Paredes S, Ong CL, Ruch P, Michel B (2014) Exergoeconomic analysis of high concentration photovoltaic thermal co-generation system for space cooling. Renew Sustain Energy Rev 34:8–19

25. Buonomano A, Calise F, Dentice d'Accadia M, Vanoli L (2013) A novel solar trigeneration system based on concentrating photovoltaic/thermal collectors. Part 1: Design and simulation model. Energy 61:59–71
26. Calise F, Dentice d'Accadia M, Piacentino A (2014) A novel solar trigeneration system integrating PVT (photovoltaic/thermal collectors) and SW (seawater) desalination: dynamic simulation and economic assessment. Energy 67:129–148
27. Meneses-Rodríguez D, Horley PP, González-Hernández J, Vorobiev YV, Gorley PN (2005) Photovoltaic solar cells performance at elevated temperatures. Sol Energy 78:243–250
28. Nishioka K, Takamoto T, Agui T, Kaneiwa M, Uraoka Y, Fuyuki T (2006) Annual output estimation of concentrator photovoltaic systems using high-efficiency InGaP/InGaAs/Ge triple-junction solar cells based on experimental solar cell's characteristics and field-test meteorological data. Sol Energy Mater Sol Cells 90:57–67
29. Torchynska TV, Polupan G (2004) High efficiency solar cells for space applications. Superficies y Vacío 17:21–25. http://www.fis.cinvestav.mx/~smcsyv/supyvac/17_3/SV1732104.PDF
30. Ton D, Tillerson J, McMahon T, Quintana M, Zweibel K (2007) Accelerated aging tests in photovoltaics summary report. U.S. Department of Energy Efficiency and Renewable Energy
31. Kosmadakis G, Manolakos D, Papadakis G (2011) Simulation and economic analysis of a CPV/thermal system coupled with an organic Rankine cycle for increased power generation. Sol Energy 85:308–324
32. Final Report DSTI—GUTS Techno Project (2013) Evaluation of the use of SolSep membranes in process industries. http://ispt.eu/cusimages/Projects/CS-01-01%20Final%20report%20Evaluation%20of%20use%20of%20Solsep%20membranes.pdf. Accessed 20 July 2013
33. Livingston A, Peeva L, Han S, Nair D, Luthra SS, White LS, Freitas Dos Santos LM (2003) Membrane separation in green chemical processing: solvent nanofiltration in liquid phase organic synthesis reactions. Ann N Y Acad Sci 984:123–141
34. Luis P, Degreve J, Van der Bruggen B (2013) Separation of methanol–n-butyl acetate mixtures by pervaporation: potential of 10 commercial membranes. J Membr Sci 429:1–12
35. Incropera FP, Dewitt DP, Bergman TL, Lavine AS (2007) Fundamentals of heat and mass transfer, 6th edn (trans: Xinshi G, Hong Y). Chemistry Industry Press, Beijing (Chinese)
36. Rabl A (1976) Optical and thermal properties of compound parabolic concentrators. Sol Energy 18:497–511
37. Kreith F (1973) Principles of heat transfer, 3rd edn. Intext Educational, New York
38. Product catalog. http://www.zeussolar.si/pdf/.22/12/2008

Chapter 3
Experimental Study of the ORC Under Variable Condensation Temperature

Combined heat and power (CHP) system provides easier market entry for the small scale organic Rankine cycle (ORC). However, in most areas heat is not continuously desirable through the year. Variable operation of the ORC is unavoidable. The condensation temperature is a key parameter. It changes with the seasonal weather condition, and influences the expansion/pressurization ratio and the temperature difference driving the ORC. The performances of the expander, the pump and the whole cycle are consequently affected. At present the information on the ORC performance under variable condensation temperature is rare. This chapter conducts a thorough experimental investigation on this subject. Special attention is paid to the practical performance of the turbo expander over a wide range of pressure ratio. The exergetic efficiency of each of the main components varying with the condensation temperature is revealed.

3.1 Introduction

CHP is important to a cost-effective ORC at low power [1, 2]. Domestic CHP generation is deemed a new application of the ORC [3]. To an ORC-based domestic CHP system, the variable operation will be unavoidable through the year due to the fluctuation of available thermal energy of the heat source and heat sink [4–8]. In particular, the variable operation with respect to the condensation temperature will be executed for most ORC-CHP systems. First, the consumer's demand on space heating or bathing varies with the seasonal weather condition. In many areas of the world, a fluctuation of ambient temperature of 40 °C is common and there is no need of space heating in summer, spring, and autumn. And the amount of hot water for bathing may also be reduced in some seasons. The exhausted heat from the condensation of the ORC fluid in hot weather conditions becomes residual, and may even cause superheat for the residential buildings. So the ORC exhausted heat is not always desirable. Second, since there is no need for the ORC to operate under high condensation temperature over some period, the cold side temperature of the ORC shall be close to the ambient temperature to achieve more power output.

J. Li, *Structural Optimization and Experimental Investigation of the Organic Rankine Cycle for Solar Thermal Power Generation*, Springer Theses,
DOI 10.1007/978-3-662-45623-1_3

The condensation temperature for the CHP operation is generally higher than 50 °C. The ORC power efficiency under this condensation temperature is theoretically lower than that under the condensation temperature from 20 to 40 °C, which can be seen by Figs. 3.1 and 3.2. The expander, the pump and the generator efficiencies are assumed to be 0.7, 0.5 and 0.85. The evaporation temperature is from 100 to 140 °C. The efficiency increases almost linearly with the decrement in the condensation temperature for both fluids. It is clear the potential increment of the power efficiency is significant when the ORC shifts from the CHP generation.

Though the theoretic analysis indicates the necessity of variable operation, the performance of a real ORC under variable condensation temperature is difficult to predict. It is convenient to assume constant expander efficiency, constant pump efficiency, etc. for the analysis on different conditions. However, the devices in the ORC are designed for certain operating conditions. To investigate the ORC performance at variable operation and to estimate the yearly power generation, it is of key importance to take into account the part load behavior of the system components [9]. At part load operation their behavior may be quite different [10, 11]. The larger the deviation of the practical condition from the design is, the lower the efficiency may be. In this case, the constant efficiency assumption seems no longer rigorous. The higher power efficiency at lower condensation temperature needs experimental validation.

The influence of the condensation temperature on the performances of the expander, the pump, etc. in the ORC is much more appreciable than that in the conventional steam Rankine cycle. The saturation pressure of water is very low at

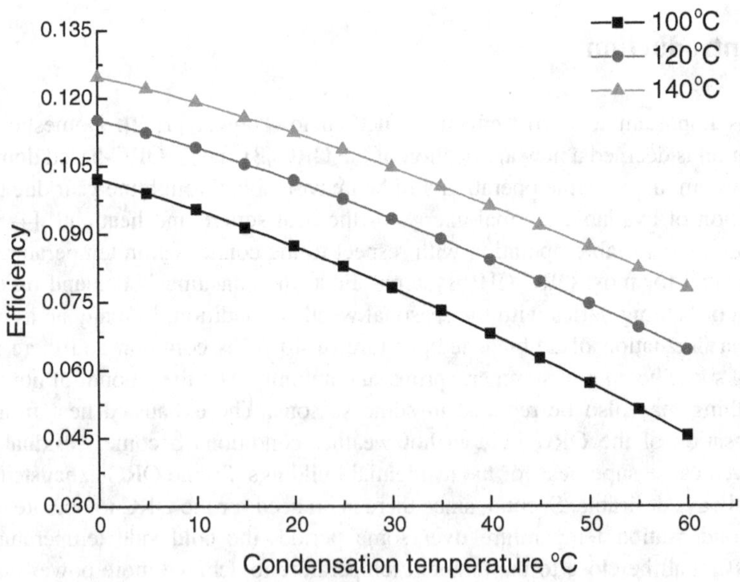

Fig. 3.1 The ORC electricity efficiency variation with the condensation temperature using R123

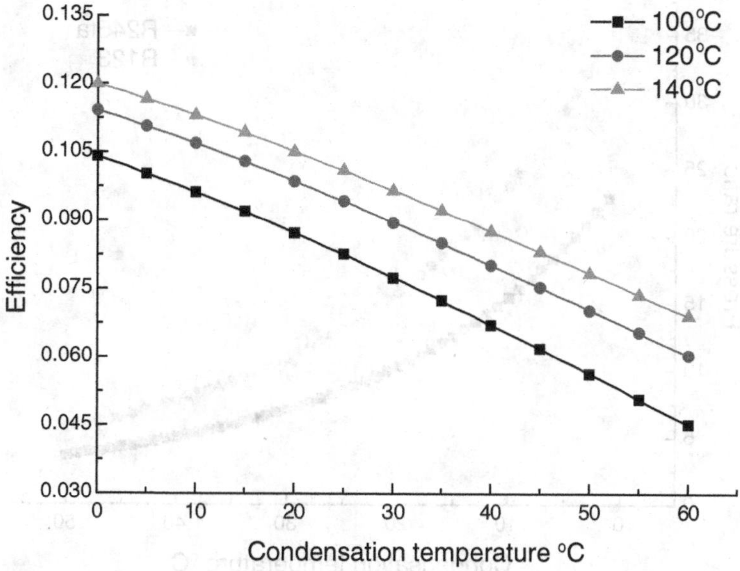

Fig. 3.2 The ORC electricity efficiency variation with the condensation temperature using R245fa

normal ambient temperature. In practical operation of the fossil fuel-fired power plant it is difficult to get an outlet pressure of steam turbine lower than 5 kPa. This technical limit for maintaining a vacuum in the steam condenser has been mentioned in previous studies [12–14]. Lower backpressure and larger enthalpy drop are unavailable for steam turbine at condensation temperature lower than 33 °C. While in the ORC this technical limit can be avoided. For example, the saturation pressures of R123 and R245fa at 33 °C are 122 and 198 kPa, respectively. The saturation pressures drop to 33 and 53 kPa respectively at a condensation temperature of 0 °C, which are still much higher than 5 kPa. Therefore, the ORC can react well to low condensation temperature. As the condensation temperature gets lower, the loads of the condenser and evaporator increase, which may enlarge the heat transfer irreversibility in the heat exchangers. Moreover, as the condensation temperature changes, the pressure ratio for the expander and the pump will change, which is shown in Fig. 3.3. The evaporation temperature for R245fa is 100 °C, while it is 120 °C for R123. As the pressure ratio changes, the enthalpy drop during expansion and the enthalpy increment during pressurization will vary accordingly. The performances of the expander, the pump and the whole cycle may seriously deviate from the design.

The pressure ratio is especially important for the turbo expander. The sound speeds of most organic fluid are much lower than that of steam, and transonic flow is the typical flow of ORC expander [15, 16]. The variation of the pressure ratio will disturb the internal flow of the expander. Shock wave or expansion wave may form and the energy conversion inside will be consequently affected. Besides, the variation of the pressure ratio leads to strong deviation of the leaving loss of the

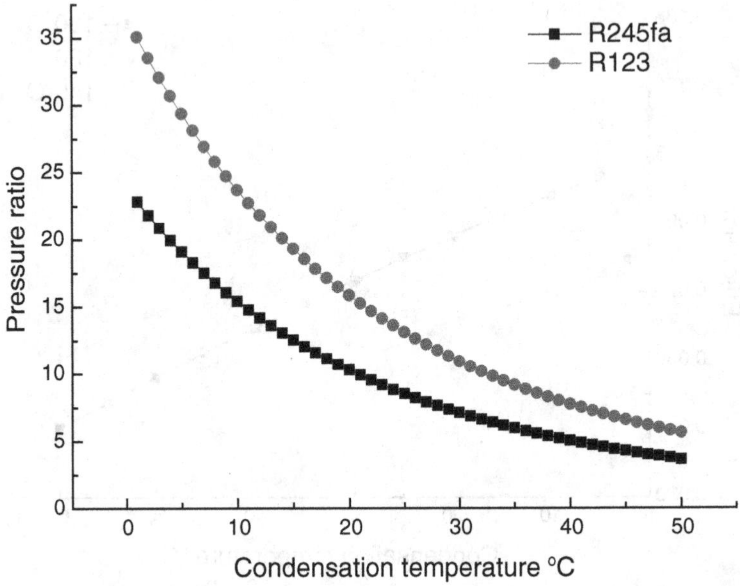

Fig. 3.3 The pressure ratio variation with the condensation temperature

expander [17]. The expander leaving loss increases with the decrement in the condensation temperature due to larger specific volume and mass flow rate. It will go up by about 10 times when the condensation temperature changes from 30 to 0 °C using the working fluids of R123 and R245fa. The leaving loss will account for an appreciable percentage of the total energy loss in the energy conversion and considerably influence the expander output power when the condensation temperature drops greatly below the design.

At present the information on the performance of the ORC expander at variable operation is limited. Zheng et al. analyzed the influence of the variable rotation speed on a rolling-piston expander [18]. The test results showed that the expander normally ran between 350 and 800 rpm. The expander mechanical efficiency varied with the rotation speed. A maximum mechanical efficiency of 44 % was achieved at 645 rpm. Lemort et al. carried out an experimental study on a prototype of an open-drive oil-free scroll expander working with R123 [19]. The expander shaft power varying with the imposed pressure ratio was measured. The expander under variable pressure ratio was modeled in a semi-empirical way. The internal leakages and, to a lesser extent, the supply pressure drop and the mechanical losses were deemed the main losses affecting the performance of the expander. Bracco et al. also carried out the experimental test and modeling of a scroll expander [20]. The registered working parameters and efficiencies were comparable with those expected from previous studies. The model revealed a good skill in predicting the main working parameters of the expander and the ORC. Avadhanulal et al. tested a 50 kW ORC under different conditions of the heat source and heat sink using a screw expander [21]. Empirical models of high accuracy were built. The predicted values of the screw expander power output were within 7.5 % compared with the experimental data. To the best knowledge of the authors, experimental investigation of the turbo expander over a wide range of pressure ratio for small scale ORC application has not been conducted yet.

As for the system, the experimental test with respect to the condensation temperature is rare. Erhart et al. introduced a commercial ORC-CHP plant with 5.3 MW_{th} and 1 MW_{el} nominal output in Scharnhauser Park, Ostfildern, Germany [22]. The combustion heat of the wood chips was transferred to the ORC via a thermal oil system with feed temperature of 300 °C and return temperature of 240 °C. The system efficiency was analyzed for the summer and winter cases respectively. And the influence of condenser conditions on ORC load characteristics was investigated. Given a relatively steady inlet temperature of cooling water, ORC power increased by 1.5 % at full load (winter case) and 1.0 % at part load (summer case) with the increment in water mass flow rate, resulting from the higher heat transfer coefficient and lower outlet temperature of water. However, the detailed behavior of the expander, pump and heat exchangers was not revealed.

All above issues imply great uncertainty of the ORC performance under variable condensation temperature. This chapter presents a thorough experimental investigation of the ORC in the condensation temperature range from about 20 to 50 °C with low-grad heat source. A specially designed turbo expander is used. The feasibility of the turbo expander and the ORC for small-scale domestic CHP generation is demonstrated.

3.2 Thermodynamic Formulas

The energetic efficiency of the ORC can be calculated by Eq. (2.7). The exergy loss associated with element *i* is the difference between the exergy E_i^{in} flowing into *i* and the exergy E_i^{out} leaving it [23, 24]:

$$\begin{aligned} \xi_i &= E_i^{\text{in}} - E_i^{\text{out}} \\ &= E_i^a - E_i^u \end{aligned} \tag{3.1}$$

E_i^a is the available exergy for *i*, and E_i^u is the used exergy. The exergy efficiency of *i* is determined by

$$\eta_{ex,i} = \frac{E_i^u}{E_i^a} \tag{3.2}$$

The elements in the ORC are mainly expander, pump, evaporator, condenser, storage tank, separator, pipes, etc.

For the expander

$$E_t^a = E_{t,i} - E_{t,o} \tag{3.3}$$

$$E_t^u = W_t \tag{3.4}$$

where $E_{t,i}$, and $E_{t,o}$ are the expander inlet exergy and outlet exergy.

For the pump

$$E_p^a = W_p \tag{3.5}$$

$$E_p^u = E_{p,o} - E_{p,i} \tag{3.6}$$

where $E_{p,i}$ and $E_{p,o}$ are the pump inlet exergy and outlet exergy.

For the evaporator

$$E_e^a = E_{h,i} - E_{h,o} \tag{3.7}$$

$$E_e^u = E_{f,o} - E_{f,i} \tag{3.8}$$

where $E_{h,i}$ and $E_{h,o}$, $E_{f,i}$ and $E_{f,o}$ are the exergies of the hot side fluid at the evaporator inlet and outlet, the exergies of the organic fluid at the evaporator inlet and outlet.

For the condenser

$$E_c^a = E_{f,i} - E_{f,o} \tag{3.9}$$

$$E_c^u = E_{w,o} - E_{w,i} \tag{3.10}$$

where $E_{w,i}$ and $E_{w,o}$ are the exergies of water at the condenser inlet and outlet.

For the pipes, separator, tank and other elements, throttling process is assumed. And therefore the used exergy is zero. The available exergy is the difference of the inlet and outlet exergy at each of the elements.

The total exergy loss of the system is the sum of the exergy losses in the all the elements:

$$\xi_{\mathrm{ORC}} = \xi_t + \xi_p + \xi_e + \xi_c + \xi_{others} \tag{3.11}$$

The system exergy efficiency is the ratio of the total used exergy to the total available exergy of the system:

$$\eta_{\mathrm{ex,ORC}} = \frac{E_{\mathrm{total}}^u}{E_{\mathrm{total}}^a} = \frac{E_t^u + E_c^u}{E_e^a + E_p^a}. \tag{3.12}$$

3.3 Experiment Setup

Figure 3.4 shows the test rig of the ORC system. The working fluid is R123. The system consists of three subcircuits: R123 circuit, conduction oil circuit, and water

Fig. 3.4 Test rig of the ORC system: *1* expander; *2* gearbox; *3* generator; *4* condenser; *5* pump; *6* evaporator; *7* separator; *8* temperature controller

circuit. For the first one, R123 is vaporized under high pressure in the evaporator, and heat is transferred from the oil to R123. The vapor flows into the expander, exporting power in the process due to enthalpy drop. The outlet vapor is condensed to liquid state in the condenser, and heat is transferred from R123 to water. The liquid is pressurized and pumped into the evaporator. R123 circulates in such a way. A separator is used to prevent the expander from being hit in case there are droplets at the evaporator outlet. A gearbox of reduction ratio 20:1 is employed to ensure a lower rotation speed of the generator. For the second circuit, the oil leaving the evaporator at lower temperature flows into the temperature controller. It is then heated. The oil at higher temperature is sent to the evaporator. The oil temperature at the outlet of the controller can be adjusted in the range from the environment temperature to 200 °C. For the third circuit, the water leaving the condenser at higher temperature flows into the cooling tower as shown in Fig. 3.5a. It is cooled down inside. Water at lower temperature is sent back to the condenser. The water outlet is at the bottom of the tower.

The condensation temperature of R123 is strongly related to the temperature of water coming from the tower. Normally, the cooling tower is a heat removal device used to transfer waste heat to the atmosphere. This tower uses the evaporation of water to remove heat and makes the outlet water of the tower be near the wet-bulb air temperature. However, by means of insulation it is possible to increase the temperature of the outlet water as shown in Fig. 3.5b. Water of relatively high temperature can be observed in Fig. 3.5c (water vapor condensed to fine particles).

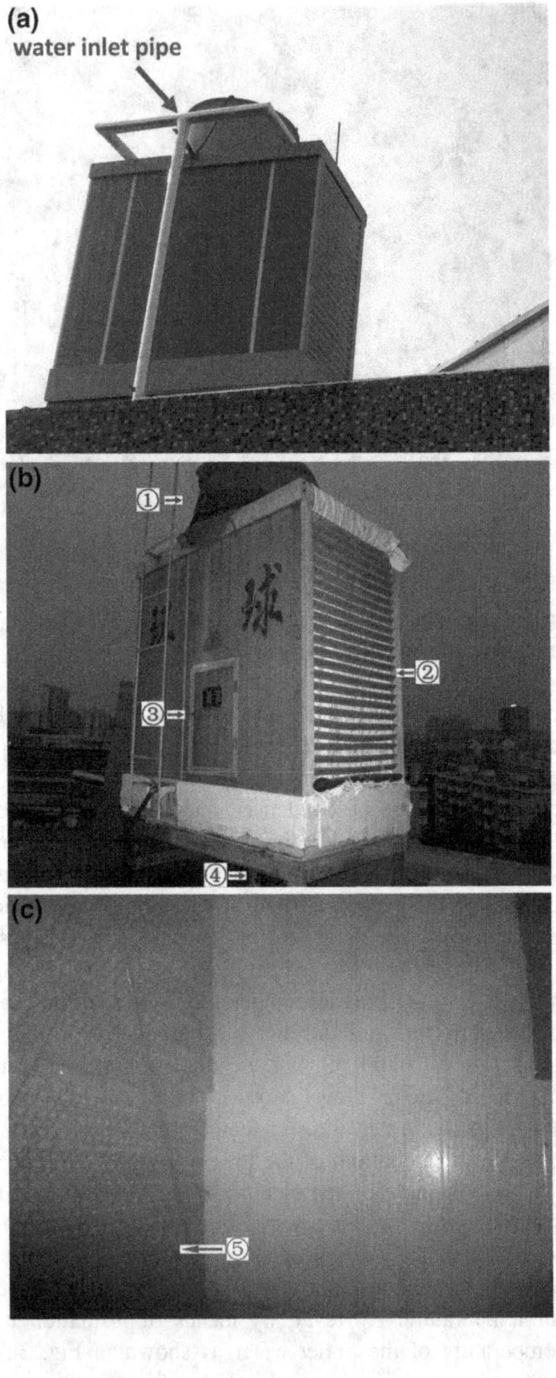

Fig. 3.5 The cooling tower **a** normal state; **b** wrapped **c** the inside. *1* Air-in by fan; *2* air-out by window; *3* door; *4* water outlet; *5* fins

A simplified structure of the system is depicted in Fig. 3.6. The line segments along with T and P in the figure are the measurement positions for temperature and pressure. Copper-constantan thermocouples are used to measure the fluid temperature, with an accuracy of ±0.5 °C. Two kinds of ceramic pressure transmitters manufactured by Huba Control Co., which range from 0 to 25 bar and from −1 to 9 bar (gage pressure), respectively, with an accuracy of ±1.0 %, are employed in the test to measure the pressure. The flow meter for R123 is MFM2081K-60P/DN25, provided by KROHNE. It is mounted near the outlet of the pump. The zero point stability and the measurement accuracy are ±0.012 kg/min and ±0.15 %MV+Cz respectively. The rotation speed is measured by the frequency signal output of the generator. All the signals are recorded and stored on a disk via Agilent Bench Link Data Logger, a computer data-acquisition system.

The oil temperature controller is AOS-50, provided by Aode Co. The maximum output heat is 100 kW. And the maximum mass flow rate of the conduction oil is 7.5 L/s.

The pump for R123 pressurization is CR1-30, provided by GRUNDFOS. This centrifugal pump is connected with a frequency converter. R123 mass flow rate can

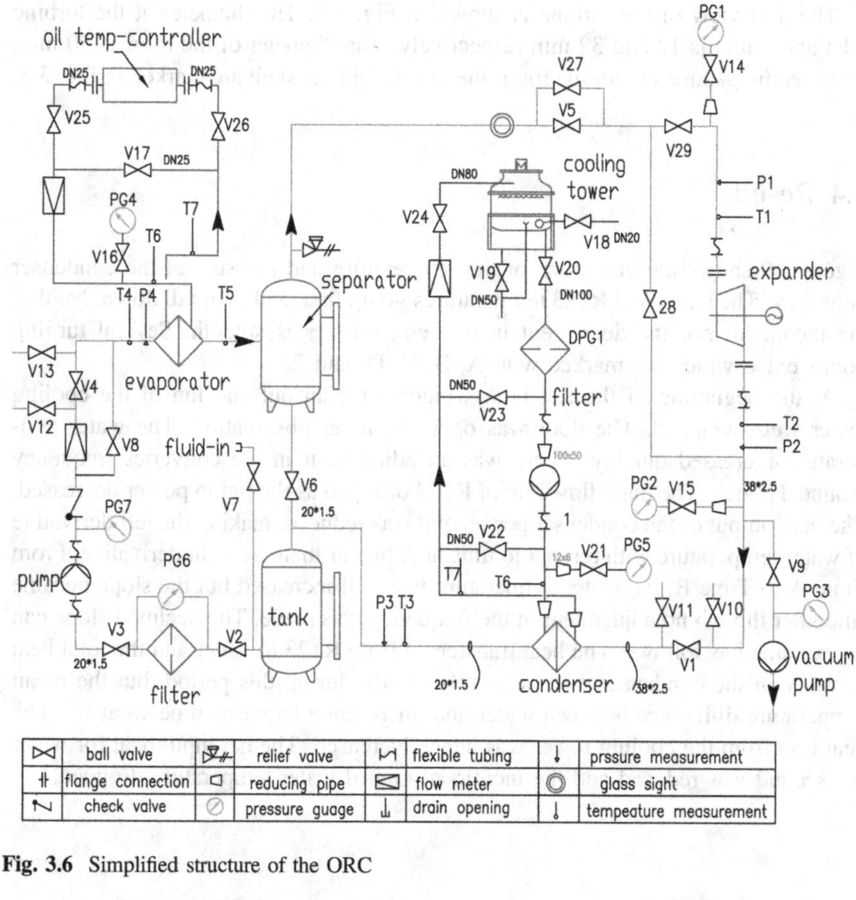

Fig. 3.6 Simplified structure of the ORC

be adjusted by the converter frequency. The height difference between the pump inlet and the fluid tank outlet is about 0.8 m.

The generator is 8SC3110, provided by Prestolite Co., which requires excitation with an input voltage of 24 V. Load to the generator can be adjusted by bulbs. This generator was originally used in an air-conditioned bus, and was not designed for the ORC application.

Plate heat exchangers provided by SWEP Co. are used in the heating and cooling processes. The evaporator is made up of about 130 plates, of a total heat transfer area of about 7.5 m^2. The condenser is made up of about 120 plates, of a total heat transfer area of about 12.0 m^2.

Among the components in this ORC, the expander is the only one that is unavailable on the market. It has been specially designed and manufactured. It is a turbo expander. For this impulse-reaction turbine, there is a stator vane and a rotor blade arrangement. The expansion of working fluid takes place not only in the nozzle but also over the blades. The blades within the chamber are located radially, pointing to the radical direction close to the nozzle and gradually turning to the axial direction. This radial-axial design provides many advantages such as a compact structure with good manufacturability, high ratio of power to volume, high efficiency, and a single-stage expansion rate of large enthalpy drop.

The 3-D view of the turbine is shown in Fig. 3.7. The diameter at the turbine inlet and outlet is 12 and 32 mm respectively. The diameter of the rotor is 50 mm. The specific parameters for the rotor, the nozzle and the shell are marked in Fig. 3.8.

3.4 Results

Figure 3.9 shows the variations of the temperature and pressure at the condenser boundary. The water and R123 temperatures go up first, and then fall down. Neither the increment nor the decrement in the temperatures is smooth. Several turning points exist, which are marked with A, B, C, D, and E.

At the beginning of the test both windows for air-out and fan of the cooling tower were wrapped. The door was open for inner observation. The water temperature increased quickly. There was an adjustment in the converter frequency around Time A. The mass flow rate of R123 dropped as the pump power decreased. The heat output of the condenser per second was reduced, making the left derivative of water temperature with respect to time at A higher than the right derivative. From Time A to Time B, the water temperature was still increased but the slope became smoother though no adjustment in the frequency was made. The declined slope can be explained as follows: The heat transferred from R123 to water and the total heat capacity in the condenser were relatively steady during this period, but the mean temperature difference between water and air became larger as time went by. The heat loss from the cooling tower was hence increased. The net input heat for water per second was reduced and the increment rate of water temperature dropped.

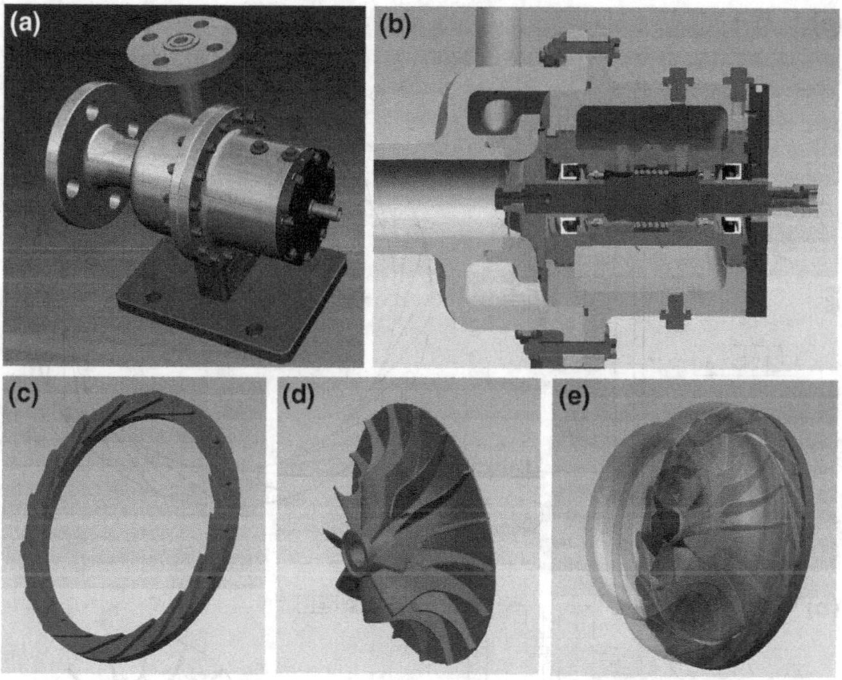

Fig. 3.7 3-D view of the turbine: **a** external; **b** internal; **c** nozzle; **d** rotor; **e** connection of the nozzle, rotor and end cover

From Time B to Time C, the tower door was close and the convective heat transfer from water to air was reduced, resulting in a quicker increment in the water temperature. From Time C to Time D, the door was open again and the fan was unwrapped. The heat and mass transfer between water and air was accelerated dramatically. The heat loss from the cooling tower became larger than the heat input from the condenser, and the water temperature fell down. From Time D to Time E, the air-out window was unwrapped, which enlarged the heat and mass transfer area, leading to a sudden decrement in the slope around Time D. As the water temperature decreased, the temperature difference between water and air got smaller and the temperature curve became smoother. Around Time E the fan was turned on. The heat and mass transfer coefficients reached the maximum.

The outlet temperature and pressure of R123 varied in a similar way to the water temperature. There was a small degree of supercool of R123 at the condenser outlet.

The R123 temperature at the evaporator inlet and outlet varying with time is shown in Fig. 3.10. Due to the oversurfacing of the evaporator, the outlet temperature of R123 was close to the inlet temperature of the conduction oil, which was assigned 105 °C. The inlet temperature of R123 for the evaporator was related to the condensation temperature. The variation of R123 mass flow rate at the evaporator inlet is also shown in Fig. 3.10. There were sudden increase and decrease around 17:16,

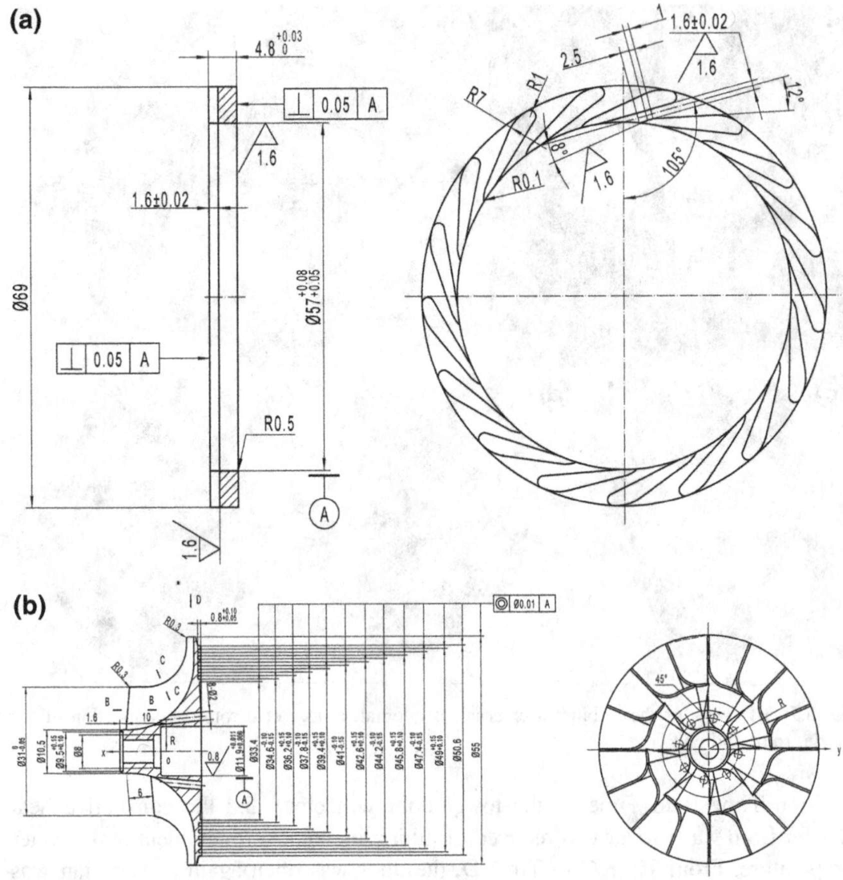

Fig. 3.8 Dimension of the nozzle and rotor: **a** nozzle; **b** rotor

18:09 and 18:28 respectively, which were caused by the adjustment of the converter frequency for pumping and the opening degree of the expander inlet valve.

Figure 3.11 shows the pump operation pressure and power varying with time. The pump power and the outlet pressure fluctuated at the startup stage due to adjustments. After 17:07 the expander inlet valve was fully open and the pump power kept relatively steady. The pump outlet pressure increased with the increment in the inlet pressure. A proof by contradiction will show that given a constant input power, the outlet pressure of the pump can not decrease or be unchanged as the inlet pressure increases in the ORC: If it could, then the mass flow rate through the pump would increase due to a lower resisting pressure difference between the pump outlet and inlet. While the mass flow rate through the expander would decrease due to a lower driving pressure difference between the expander inlet and outlet. The mass flow rate through the pump would be higher than that through the

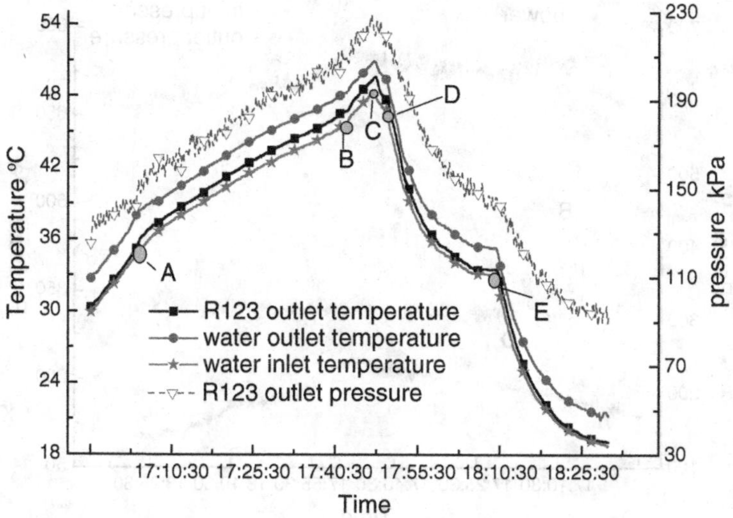

Fig. 3.9 The variations of water *inlet* and *outlet* temperature, R123 condensation temperature and pressure with time

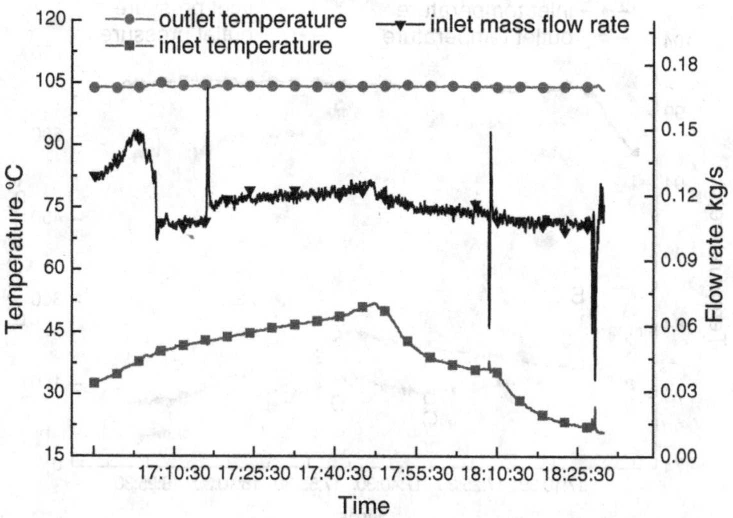

Fig. 3.10 The variations of R123 temperature and mass flow rate for the evaporator

expander, and the mass balance in the ORC would be broken. This surely is not the case in a quasi-steady state.

Figure 3.12 shows the parameters for the expander varying with time. The inlet temperature of R123 was about 101 °C, a little lower than the temperature of R123

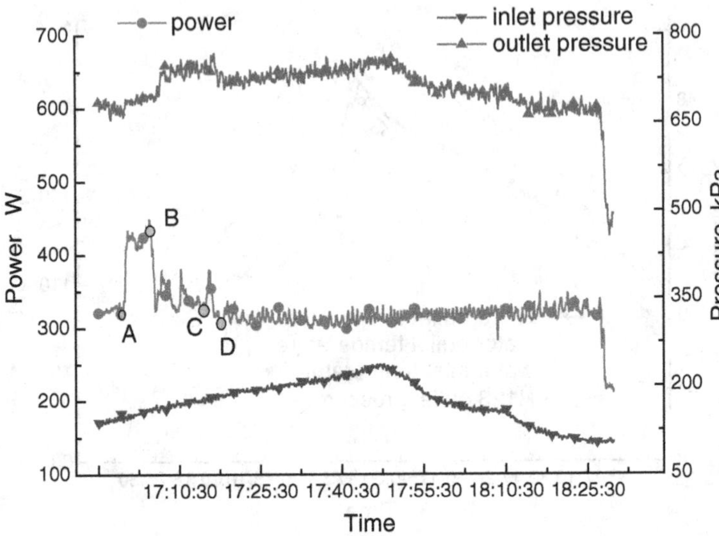

Fig. 3.11 The variations of the pump power and R123 *inlet* and *outlet* pressures

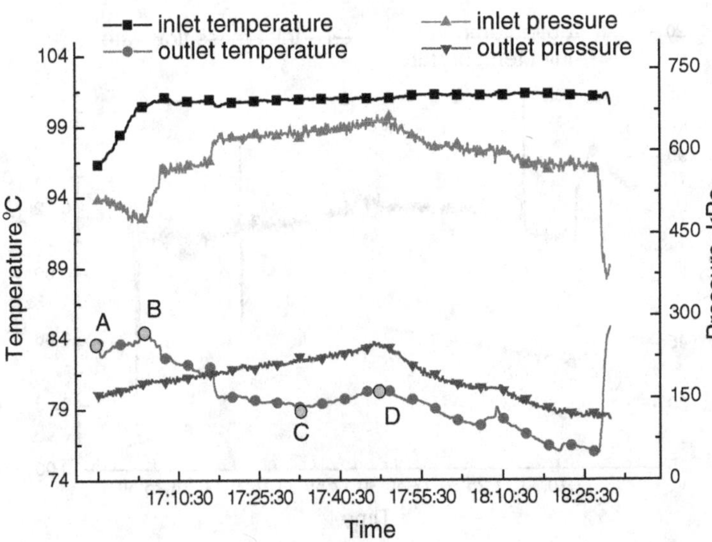

Fig. 3.12 The variations of R123 temperature and pressure at the expander *inlet* and *outlet*

leaving the evaporator due to the heat loss. The outlet temperature fluctuated more strongly than the inlet temperature. The outlet temperature was influenced by the pressure ratio and the rotation speed of the expander. From Time A to Time B it increased with the decrement in the expansion ratio. Though the rotation speed was up in this period, its influence on the outlet temperature was secondary. From Time B to Time C both the rotation speed and the inlet pressure of the expander increased, resulting in the decreased outlet temperature. From Time C to Time D the outlet temperature increased again with the increment in the outlet pressure. There was no significant increment in the inlet pressure and the rotation speed. From Time D it decreased gradually with the decrement in the inlet and outlet pressures.

The inlet pressure of the expander was correlated with the outlet pressure of the pump. However, at the beginning of the experiment, it varied differently from the pump outlet pressure due to the adjustment of the inlet valve of the expander.

Figure 3.13 shows the generator rotation speed, input shaft power and output electricity varying with time. The input shaft power for the generator set consisting of the gearbox was defined as the output shaft power of the expander. The generated electricity increased with the increment in the rotation speed at the beginning, indicating that higher rotation speed led to more adequate expansion of R123. However, from 17:16, this trend stopped due to the trade-off between the increment in the expander efficiency and the decrement in the driving temperature difference of the ORC. After about 17:32, the rotation speed of the generator fluctuated around 1,150 rpm, with a corresponding expander rotation speed of about 23,000 rpm. Then the condensation temperature and pressure became the primary factor influencing the electricity output.

Figure 3.14 shows the efficiency variations of the pump, generator set and expander. The efficiency of the generator set was defined by the ratio of the electricity generated to the expander output shaft power, and thus the gearbox efficiency was comprised. The pump and the generator set efficiencies were about 0.14 and 0.3 respectively. The efficiency of the pump was low because this kind of pump was

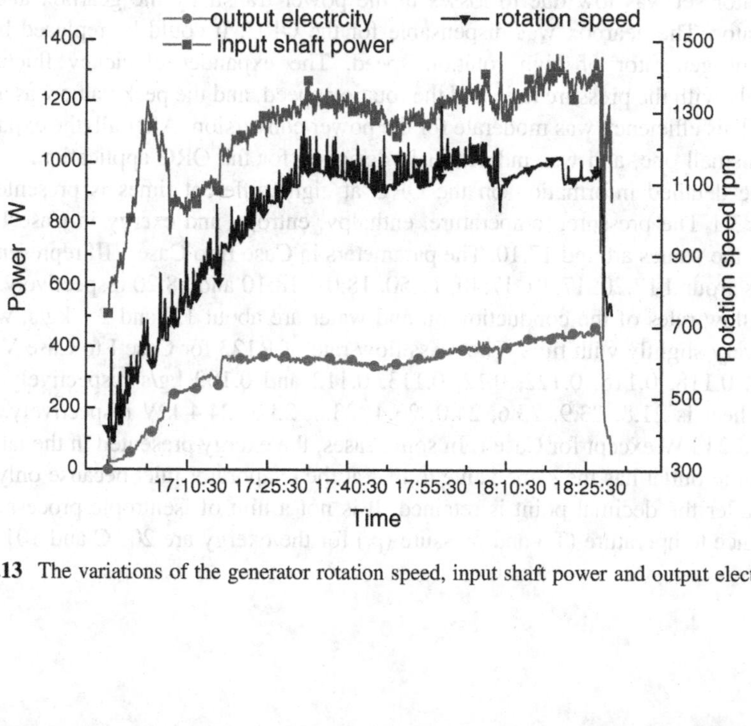

Fig. 3.13 The variations of the generator rotation speed, input shaft power and output electricity

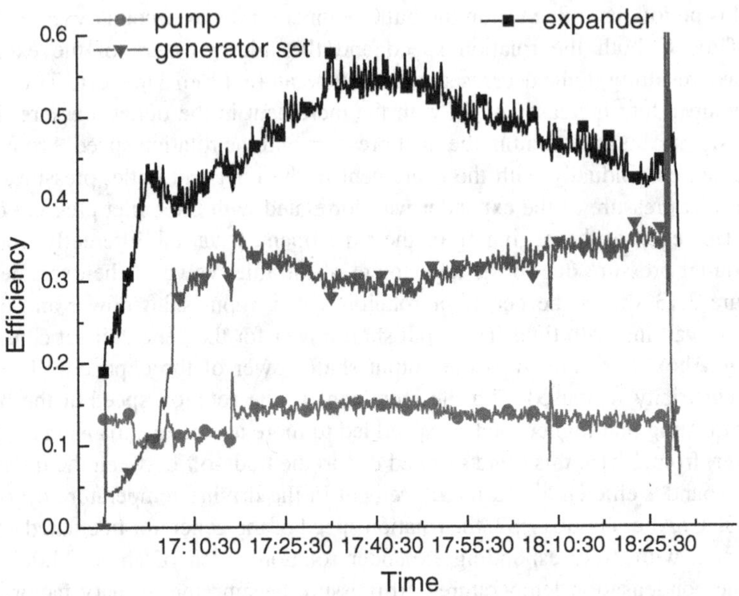

Fig. 3.14 The variations of the expander, pump, and generator efficiencies

generally used for water pressurization. When the working fluid was R123, the vapor pressure ahead of the pump became much higher and cavitation was more easily facilitated. And the nominal mass flow rate of this pump was 0.32 kg/s, indicating that it was at off-design operation in the test. The efficiency of the generator set was low due to losses in the power transit by the gearbox and the generator. The gearbox was dispensable for the ORC. It could be replaced by an efficient generator of high rotation speed. The expander efficiency fluctuated strongly with the pressure ratio and the rotation speed, and the peak value was about 0.54. This efficiency was moderate for the power conversion. After all, the expander was a small one, and was indigenously designed for the ORC application.

The detailed information on the ORC at eight different times is presented in Table 3.1 The pressure, temperature, enthalpy, entropy and exergy in Case I represent the values around 17:10. The parameters in Case II to Case VIII represent the values around 17:20, 17:30, 17:40, 17:50, 18:00, 18:10 and 18:20 respectively. The mass flow rates of the conduction oil and water are about 1.8 and 2.1 kg/s, which vary very slightly with time. The mass flow rate of R123 for Case I to Case VIII is 0.106, 0.118, 0.118, 0.122, 0.12, 0.113, 0.112 and 0.109 kg/s respectively. The input heat is 21.8, 23.9, 23.6, 24.0, 23.4, 23.6, 23.6, 24.4 kW respectively. It is around 24 kW except for Case I. In some cases, the exergy presented in the table at the pump outlet has the same value to that at the evaporator inlet because only one digit after the decimal point is retained. It is not a hint of isentropic process. The reference temperature (T_r) and pressure (p_r) for the exergy are 20 °C and 101 kPa.

Table 3.1 Specific thermodynamic parameter distributions on the ORC

	State point	Fluid	*p* (kPa)	*t* (°C)	*h* (kJ/kg)	*s* (J/kg/°C)	*e* (kJ/kg)
Case I	Expander inlet	R123	569	100.7	444.4	1,713	36.0
	Expander outlet	R123	182	82.3	436.4	1,749	17.5
	Pump inlet	R123	171	38.0	238.5	1,132	0.5
	Pump outlet	R123	664	40.8	241.6	1,140	1.2
	Evaporator inlet	R123	635	41.0	241.8	1,141	1.2
	Evaporator outlet	R123	597	104.0	446.7	1,717	37.2
	Controller inlet	Oil	–	100.1	219.5	685	31.2
	Controller outlet	Oil	–	105.0	231.0	716	33.7
	Condenser inlet	R123	180	82.0	436.2	1,749	17.3
	Condenser outlet	R123	162	38.0	238.5	1,132	0.5
	Tower inlet	Water	–	39.8	166.8	570	2.7
	Tower outlet	Water	–	37.3	156.3	536	2.1
Case II	Expander inlet	R123	624	100.7	443.4	1,706	37.1
	Expander outlet	R123	201	79.9	434.2	1,738	18.5
	Pump inlet	R123	188	41.0	241.7	1,142	0.7
	Pump outlet	R123	730	43.4	244.3	1,149	1.3
	Evaporator inlet	R123	695	43.5	244.4	1,149	1.4
	Evaporator outlet	R123	669	104.3	445.7	1,709	38.5
	Controller inlet	Oil	–	99.3	217.8	680.6	30.9
	Controller outlet	Oil	–	105.0	231.0	716	33.7
	Condenser inlet	R123	177	79.6	434.4	1,745	16.7
	Condenser outlet	R123	175	41.0	241.6	1,142	0.6
	Tower inlet	Water	–	42.8	179.3	610	3.5
	Tower outlet	Water	–	40.2	168.5	575	2.8
Case III	Expander inlet	R123	622	100.9	443.5	1,706	37.2
	Expander outlet	R123	212	79.5	433.7	1,733	19.5
	Pump inlet	R123	201	43.7	244.5	1,151	0.9
	Pump outlet	R123	740	46.2	247.2	1,158	1.5
	Evaporator inlet	R123	698	46.7	247.7	1,160	1.5
	Evaporator outlet	R123	664	104.3	445.8	1,710	38.3
	Controller inlet	Oil	–	99.4	218.0	681	30.9
	Controller outlet	Oil	–	105.0	231.0	716	33.7
	Condenser inlet	R123	200	80.0	434.3	1,738	18.6
	Condenser outlet	R123	192	43.7	244.5	1,151	0.9
	Tower inlet	Water	–	45.4	190.2	644	4.4
	Tower outlet	Water	–	42.8	179.3	610	3.5

(continued)

Table 3.1 (continued)

	State point	Fluid	p (kPa)	t (°C)	h (kJ/kg)	s (J/kg/°C)	e (kJ/kg)
Case IV	Expander inlet	R123	642	101.0	443.4	1,705	37.4
	Expander outlet	R123	227	79.7	433.6	1,730	20.3
	Pump inlet	R123	213	46.0	246.8	1,158	1.1
	Pump outlet	R123	735	48.2	249.3	1,165	1.6
	Evaporator inlet	R123	701	48.5	249.6	1,166	1.6
	Evaporator outlet	R123	677	104.2	445.5	1,708	38.6
	Controller inlet	Oil	–	99.3	217.7	680	30.9
	Controller outlet	Oil	–	105.0	231.0	716	33.7
	Condenser inlet	R123	220	79.5	433.6	1,731	20.0
	Condenser outlet	R123	205	46.0	246.9	1,158	1.2
	Tower inlet	Water	–	47.5	198.97	671	5.1
	Tower outlet	Water	–	44.9	188.1	637	4.2
Case V	Expander inlet	R123	647	101.0	443.3	1,705	37.3
	Expander outlet	R123	239	80.2	433.8	1,727	21.3
	Pump inlet	R123	229	47.5	248.4	1,163	1.3
	Pump outlet	R123	755	49.8	251.0	1,170	1.8
	Evaporator inlet	R123	713	50.2	251.4	1,171	1.9
	Evaporator outlet	R123	685	104.2	445.3	1,707	38.7
	Controller inlet	Oil	–	99.4	218.1	681	30.9
	Controller outlet	Oil	–	105.0	231.0	716	33.7
	Condenser inlet	R123	230	80.0	433.8	1,730	20.5
	Condenser outlet	R123	220	47.5	248.4	1,163	1.3
	Tower inlet	Water	–	49.0	205.2	691	5.6
	Tower outlet	Water	–	46.3	194.0	656	4.7
Case VI	Expander inlet	R123	611	101.2	444.1	1,709	36.9
	Expander outlet	R123	179	78.5	433.5	1,742	16.6
	Pump inlet	R123	170	35.4	235.8	1,123	0.4
	Pump outlet	R123	702	37.9	238.6	1,131	0.9
	Evaporator inlet	R123	644	37.9	238.6	1,131	0.9
	Evaporator outlet	R123	619	104.2	446.5	1,715	37.6
	Controller inlet	Oil	–	99.3	217.9	681	30.9
	Controller outlet	Oil	–	105.0	231.0	716	33.7
	Condenser inlet	R123	166	78.6	433.8	1,747	15.5
	Condenser outlet	R123	159	35.4	235.8	1,123	0.4
	Tower inlet	Water	–	37.2	155.9	535	2.0
	Tower outlet	Water	–	34.9	146.3	504	1.5

(continued)

Table 3.1 (continued)

	State point	Fluid	*p* (kPa)	*t* (°C)	*h* (kJ/kg)	*s* (J/kg/°C)	*e* (kJ/kg)
Case VII	Expander inlet	R123	594	101.3	444.5	1,712	36.4
	Expander outlet	R123	170	78.6	433.7	1,745	16.0
	Pump inlet	R123	155	33.4	233.7	1,116	0.4
	Pump outlet	R123	691	35.9	236.6	1,124	0.9
	Evaporator inlet	R123	648	36.0	236.6	1,124	0.9
	Evaporator outlet	R123	625	104.2	446.4	1,714	37.8
	Controller inlet	Oil	–	99.5	217.9	681	30.9
	Controller outlet	Oil	–	105.0	231.0	716	33.7
	Condenser inlet	R123	155	78.3	433.8	1,750	14.6
	Condenser outlet	R123	145	33.4	233.8	1,116	0.5
	Tower inlet	Water	–	35.0	146.7	505	1.6
	Tower outlet	Water	–	32.6	136.7	472	1.1
Case VIII	Expander inlet	R123	571	101.2	444.8	1,715	35.9
	Expander outlet	R123	127	76.2	432.7	1,757	11.5
	Pump inlet	R123	115	21.6	221.7	1,076	0.1
	Pump outlet	R123	674	24.2	224.6	1,085	0.3
	Evaporator inlet	R123	630	24.5	224.8	1,086	0.3
	Evaporator outlet	R123	611	104.3	446.7	1,716	37.5
	Controller inlet	Oil	–	99.2	217.6	680	30.8
	Controller outlet	Oil	–	105.0	231.0	716	33.7
	Condenser inlet	R123	115	76.1	432.8	1,763	9.8
	Condenser outlet	R123	107	21.6	221.7	1,076	0.1
	Tower inlet	Water	–	23.8	99.9	350	0.1
	Tower outlet	Water	–	21.2	89.0	314	0.0

The pressure for the conduction oil and water has not been measured. However, at liquid state the enthalpy and the entropy of the oil and water are mainly related with the temperature. For example, the specific enthalpy of water at 40 °C/0.1 MPa, 40 °C/0.2 MPa, 38 °C/0.1 MPa and 38 °C/0.2 MPa is 167.6, 167.7, 159.3 and 159.4 kJ/kg respectively.

With the data in Table 3.1, the energetic and exergetic efficiencies of the system, and the exergy losses for the main components can be calculated, as presented in Table 3.2.

The water inlet temperature $T_{w,i}$, the system input heat Q, the system power efficiency $\eta_{en,ORC}$ based on the shaft power of the expander, and the system exergetic efficiency $\eta_{ex,ORC}$ are also listed for each case. For most cases, $\eta_{en,ORC}$ decreases with the increment in $T_{w,i}$. The maximum $\eta_{en,ORC}$ is 4.1 % at $T_{w,i}$ = 21.2 °C. The relative decrement of $\eta_{en,ORC}$ is 15 % when $T_{w,i}$ climbs to 46.3 °C. On the other hand, $\eta_{ex,ORC}$ increases dramatically with the increment in $T_{w,i}$. When $T_{w,i}$ climbs from 21.2 to 46.3 °C, $\eta_{ex,ORC}$ increases from 24.1 to 57.8 %. The relative

Table 3.2 The energetic and exergetic efficiencies

		Evaporator	Expander	Condenser	Pump	Others
Case I						
$T_{w,i}$ = 37.3 °C	Available exergy (W)	4,503	1,967	1,784	329	
Q = 21.8 kW	Used exergy (W)	3,819	848	1,244	80	
$\eta_{en,ORC}$ = 2.4 %	Exergy loss (W)	684	1,119	539	249	148
$\eta_{ex,ORC}$ = 43.3 %	Exergy efficiency	84.8	43.1	69.8	24.3	
Case II						
$T_{w,i}$ = 40.2 °C	Available exergy (W)	5,099	2,193	1,892	307	
Q = 23.9 kW	Used exergy (W)	4,382	1,086	1,548	65	
$\eta_{en,ORC}$ = 3.3 %	Exergy loss (W)	716	1,107	346	242	363
$\eta_{ex,ORC}$ = 49.0 %	Exergy efficiency	85.9	49.5	82.7	21.1	
Case III						
$T_{w,i}$ = 42.8 °C	Available exergy (W)	5,018	2,090	2,091	319	
Q = 23.6 kW	Used exergy (W)	4,350	1,156	1,703	76	
$\eta_{en,ORC}$ = 3.6 %	Exergy loss (W)	668	934	388	242	245
$\eta_{ex,ORC}$ = 53.6 %	Exergy efficiency	86.7	55.3	81.4	24.0	
Case IV						
$T_{w,i}$ = 44.9 °C	Available exergy (W)	5,130	2,090	2,284	305	
Q = 24.0 kW	Used exergy (W)	4,516	1,196	1,863	55	
$\eta_{en,ORC}$ = 3.7 %	Exergy loss (W)	614	894	422	250	196
$\eta_{ex,ORC}$ = 56.3 %	Exergy efficiency	88.0	57.2	81.5	17.9	
Case V						
$T_{w,i}$ = 46.3 °C	Available exergy (W)	4,995	1,914	2,302	312	
Q = 23.4 kW	Used exergy (W)	4,413	1,138	1,929	66	
$\eta_{en,ORC}$ = 3.5 %	Exergy loss (W)	582	776	373	246	262
$\eta_{ex,ORC}$ = 57.8 %	Exergy efficiency	88.3	59.4	83.8	21.1	
Case VI						
$T_{w,i}$ = 34.9 °C	Available exergy (W)	5,043	2,291	1,703	316	
Q = 23.6 kW	Used exergy (W)	4,147	1,198	1,151	51	
$\eta_{en,ORC}$ = 3.7 %	Exergy loss (W)	896	1,093	553	265	204
$\eta_{ex,ORC}$ = 43.8 %	Exergy efficiency	82.2	52.3	67.6	16.2	
Case VII						
$T_{w,i}$ = 32.6 °C	Available exergy (W)	5,044	2,293	1,584	325	
Q = 23.6 kW	Used exergy (W)	4,126	1,210	990	62	
$\eta_{en,ORC}$ = 3.7 %	Exergy loss (W)	918	1,083	594	263	311
$\eta_{ex,ORC}$ = 41.0 %	Exergy efficiency	81.8	52.7	62.5	19.1	
Case VIII						
$T_{w,i}$ = 21.2 °C	Available exergy (W)	5,191	2,661	1,058	316	
Q = 24.4 kW	Used exergy, W	4,056	1,319	10	29	
$\eta_{en,ORC}$ = 4.1 %	Exergy loss (W)	1,135	1,342	1,048	288	366
$\eta_{ex,ORC}$ = 24.1 %	Exergy efficiency	78.1	49.6	0.9	9.0	

increment is about 140 %. The results indicate that as the condensation temperature increases, the ORC reaches a higher degree of thermodynamic perfection.

The evaporator, expander, condenser and pump are the indispensable elements in the ORC. The exergy losses in these elements predominate the total system exergy destruction. There is also thermodynamic irreversibility in the separator, flow meter, valve, pipes, etc. And the exergy destruction due to flow resistance and pressure loss in these assistant components amounts to 6–12 % of the system exergy destruction.

The available exergy for the evaporator is approximately constant except for Case I. With an available exergy of about 5,100 W, the exergy efficiency of the evaporator increases and the exergy destruction decreases with the increment in $T_{w,i}$. Given the relatively steady hot side temperatures, the sensible heat of R123 from supercool state to saturation liquid state in the evaporator decreases as the condensation temperature increases. The process of R123 in the evaporator gets closer to isothermal heating, which results in a smaller average temperature difference between the oil and R123 and a lower thermodynamic irreversibility. The exergy destruction in the evaporator amounts to about 25–30 % of the total system exergy destruction.

The available exergy for the condenser increases as $T_{w,i}$ increases. The exergy destruction in the condenser amounts to about 13–25 % of the system exergy destruction. This part of exergy destruction is mainly due to the highly superheated vapor of R123 entering the condenser. Unlike the exergy destruction in the evaporator, it is not a monotone decreasing function of the $T_{w,i}$ because the vapor temperature of R123 at the condenser inlet fluctuates with $T_{w,i}$.

With an input heat of the ORC approximating 24 kW, the heat output from the condenser is relatively steady. And the water temperature increment through the condenser is around 2.5 °C. Given an approximately constant temperature increment, the used exergy of water goes up as $T_{w,i}$ increases. A close analysis reveals that the used exergy is almost proportional to the difference between the water inlet temperature and the reference temperature ($T_{w,i} - T_r$). This result can be verified as follows.

The used exergy of water through the condenser is expressed by

$$\begin{aligned} E_c^u &= H_o - H_i - T_r(S_o - S_i) \\ &\approx m_w C_p (T_o - T_i - T_r \ln \frac{T_o}{T_i}) \end{aligned} \tag{3.13}$$

where H_i, H_0, S_i, S_o are the inlet enthalpy, outlet enthalpy, inlet entropy and outlet entropy respectively. T_r has a unit of K.

With $\Delta T = T_o - T_i$ ($\frac{\Delta T}{T_i} \ll 1$), a Taylor series can be done by

$$
\begin{aligned}
\ln\frac{T_0}{T_i} &= \ln\left(1+\frac{\Delta T}{T_i}\right) \\
&= \frac{\Delta T}{T_i} - \frac{1}{2}\left(\frac{\Delta T}{T_i}\right)^2 + \frac{1}{3}\left(\frac{\Delta T}{T_i}\right)^3 + \cdots \\
&= \frac{\Delta T}{T_i} + \mathrm{O}\left(\frac{\Delta T}{T_i}\right)
\end{aligned} \tag{3.14}
$$

Then

$$
\begin{aligned}
E_c^u &\approx m_w C_p\left(\Delta T - T_r\frac{\Delta T}{T_i}\right) \\
&\approx m_w C_p \Delta T\left(1-\frac{T_r}{T_i}\right)
\end{aligned} \tag{3.15}
$$

with $x = T_i - T_r$ $(x \ll T_r)$

$$
\begin{aligned}
E_c^u &\approx m_w C_p \Delta T\frac{x}{T_r + x} \\
&\approx \frac{m_w C_p \Delta T}{T_r}x
\end{aligned} \tag{3.16}
$$

ΔT is almost constant with the relatively steady output heat from the condenser. Therefore, E_c^u is approximately proportional to x.

The available exergy for the pump is small and equal to the input electrical power. However, the exergy destruction in the pump amounts to 7–11 % of the system exergy destruction due to the low exergetic efficiency for pumping. With the increment in $T_{w,i}$, this proportion increases and the influence of the pump on the ORC performance becomes stronger. The results in Fig. 3.14 and Table 3.2 indicate the exergetic and isentropic efficiencies of the pump are different. A relationship between the pump exergetic efficiency and its isentropic efficiency is implied. This relationship can be deduced as follow.

The specific liquid volume of R123 through the pump is almost constant. It is valid to assume that the pumping process is isovolumic. The exergy efficiency of the pump is then expressed by

$$
\begin{aligned}
\eta_{ex,p} &= \frac{E_p^u}{E_p^a} \\
&= \frac{m[(h_o - h_i) - T_r(s_o - s_i)]}{W_e} \\
&= \frac{m}{W_e}\int_i^o dh - T_r ds \\
&= \frac{m}{W_e}\int_i^o Tds + vdp - T_r ds \\
&= \frac{m}{W_e}\int_i^o vdp + \frac{m}{W_e}\int_i^o (T - T_r)ds \\
&= \frac{m}{W_e}\int_i^o vdp + \frac{m}{W_e}\int_i^o (T - T_r)c_v\frac{dT}{T}
\end{aligned}
\tag{3.17}
$$

The first item on the right is the isentropic efficiency of the pump. Since the reference temperature (20 °C) is lower than the operation temperature of the pump, the second item is positive. And the two items have the same order of magnitude: $O[v\Delta p] \approx 10^{-3} \times 10^5 = 10^2, O[(T - T_r)c_v\frac{\Delta T}{T}] \approx 10 \times 10^3 \times 1/10^2 = 10^2$. Therefore, the exergetic efficiency of the pump is higher than its isentropic efficiency.

The available exergy for the expander varies from 1,967 to 2,661 W. For most of the cases, it decreases as $T_{w,i}$ increases. The exergy destruction also decreases with the increment in $T_{w,i}$ due to a lower entropy generation during expansion. The exergy destruction in the expander amounts to about 32–41 % of the total exergy destruction, which is the largest among all the component losses. For most cases, the proportion increases with the increment in $T_{w,i}$. The exergy efficiency of the expander varies from 43 to 59 %, and is a little higher than the corresponding isentropic efficiency.

The expander exergetic and isentropic efficiencies variations with the pressure ratio are of special interest. In some previous studies, the ORC-based CHP system using turbo expander was proven to operate successfully under small temperature difference between the heat source and the cold source [25–27]. However, the detailed performance of the turbo expander was not revealed. A critical issue with the low pressure ratio expander has been raised: At low pressure ratio the specific enthalpy drop through the expander is low. The enthalpy drop may be countervailed by the admission losses, leakage losses, friction losses, etc. of the expander, resulting in an inappreciable power output. It is still unknown whether the expander losses play a more important role on the power conversion, especially when the pressure ratio gets lower.

This issue can be clarified by the results from Figs. 3.15 and 3.16. The energetic and exergetic efficiencies increase almost linearly as the pressure ratio decreases. No evidence of stronger influence of the losses on the expander performance at low

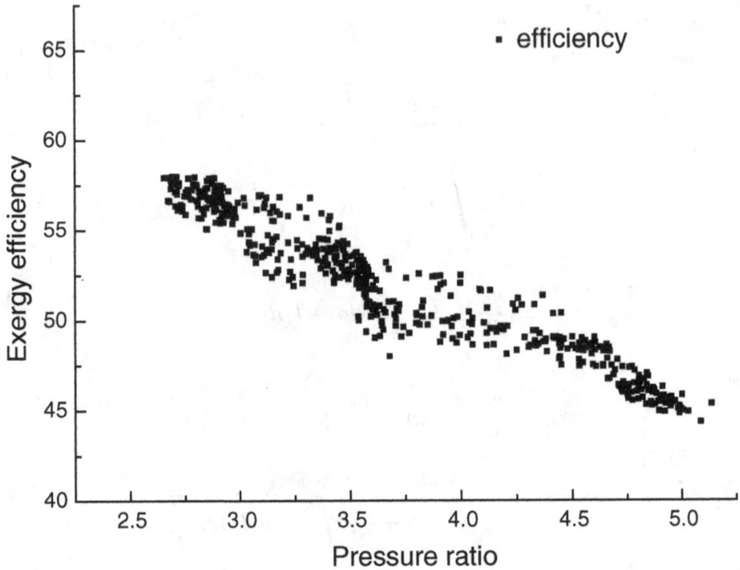

Fig. 3.15 The exergy efficiency of the expander variation with the ratio of the inlet pressure and outlet pressure

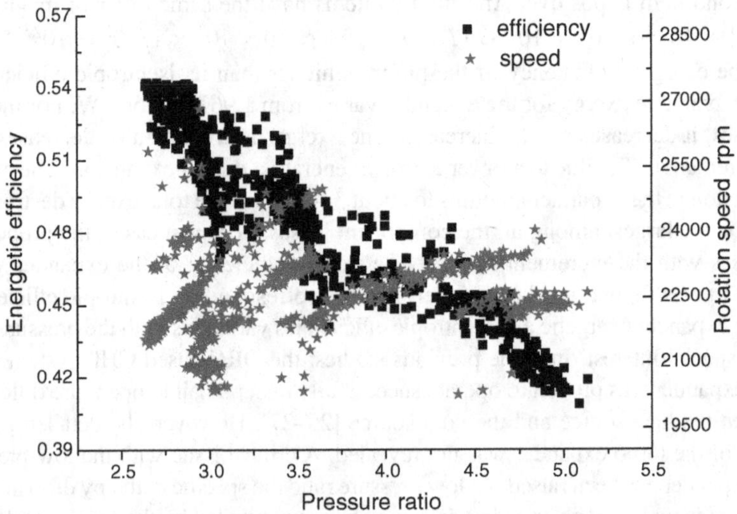

Fig. 3.16 The expander isentropic efficiency variation with the ratio of the inlet pressure and outlet pressure

pressure ratio has been found. Moreover, the results provide a novel demonstration of the feasibility of the small scale ORC-based CHP generation with low-grade heat source. The demonstration is conducted by comparing the ORC performance under variable condensation temperature:

Table 3.3 The saturation pressure of some commonly used fluids

Fluid	R123	R245fa	Butane	Pentane	Ammonia
P_s at 50 °C (kPa)	212	344	496	159	2,034
P_s at 20 °C (kPa)	76	123	208	57	857

Table 3.4 The pressure ratio and maximum Mach number of the turbo expander on different conditions

Evaporation temperature (°C)	Condensation temperature (°C)	R123		R245fa		Butane	
		Pressure ratio	Max. Ma	Pressure ratio	Max. Ma	Pressure ratio	Max. Ma
100	1	22.993	1.672	22.830	1.391	14.235	1.384
	12	14.295	1.225	14.183	1.017	9.591	1.056
	13	13.720	1.192	13.612	0.989	9.269	1.031
	14	13.173	1.160	13.068	0.962	8.960	1.006
	15	12.651	1.128	12.551	0.936	8.663	0.983
	19	10.800	1.013	10.713	0.839	7.590	0.894
	20	10.389	0.986	10.306	0.817	7.348	0.873
	30	7.169	0.757	7.113	0.626	5.384	0.691
	40	5.085	0.583	5.049	0.482	4.032	0.548
	50	3.697	0.449	3.674	0.371	3.078	0.433
120	1	35.095	1.793	34.798	1.484	20.647	1.476
	15	19.311	1.226	19.130	1.011	12.564	1.062
	16	18.553	1.194	18.378	0.985	12.151	1.038
	17	17.830	1.164	17.662	0.959	11.755	1.014
	18	17.141	1.134	16.979	0.934	11.374	0.992
	22	14.689	1.023	14.550	0.843	9.996	0.907
	23	14.143	0.998	14.011	0.821	9.684	0.887
	30	10.942	0.838	10.841	0.689	7.809	0.760
	40	7.762	0.657	7.696	0.540	5.847	0.613
	50	5.643	0.518	5.600	0.425	4.464	0.496
140	1	51.399	1.888	51.068	1.543	29.081	1.534
	17	26.113	1.237	25.920	1.007	16.557	1.064
	18	25.104	1.206	24.918	0.981	16.021	1.041
	19	24.142	1.176	23.964	0.957	15.506	1.018
	20	23.224	1.148	23.054	0.933	15.012	0.996
	25	19.221	1.015	19.081	0.824	12.813	0.895
	26	18.524	0.990	18.388	0.804	12.422	0.876
	30	16.025	0.900	15.910	0.730	10.999	0.805
	40	11.368	0.713	11.294	0.578	8.236	0.656
	50	8.265	0.569	8.219	0.461	6.288	0.537

(i) According to the experiment results, there seems to be no performance degradation of the turbo expander with regard to the internal losses at low pressure ratio. Compared with the solo power generation, the CHP generation (of relatively higher condensation temperature) can provide better thermodynamic performance for the expander. This advantage of low pressure ratio offered by the CHP generation also exists for volumetric expanders such as reciprocating expander, scroll expander, and screw expander, etc. Lower pressure ratio can reduce the leakage loss, which is an important factor affecting the efficiency of volumetric expanders. Efficient volumetric expanders generally operated under a pressure ratio lower than 6.0 [28, 29].

(ii) Efficient operation at low pressure ratio implies easier design of the expander. The outlet pressure of the expander is related to the condensation temperature. For lots of organic fluids, the saturation pressure at 50 °C is about 3 times that at 20 °C as shown in Table 3.3. It means at the same evaporation temperature, the pressure ratio of the expander operating in a CHP system may be only about 1/3 of that in a single-functional system for power generation. With a lower pressure ratio the specific enthalpy drop during expansion is reduced. Multistage expansion can be therefore avoided, offering lower cost and simpler structure of the expander.

More data by theoretical analysis is provided in Table 3.4. The evaporation temperature is 100, 120 and 140 °C respectively. The expander inlet and outlet pressures are equal to the fluid saturation pressures at the evaporation and condensation temperature respectively. The maximum Mach number denotes the ratio of the maximum available velocity in the expander to the local sound speed, which can be achieved by an isentropic expansion at the nozzle outlet with zero degree of reaction. For all the fluids, the Mach number is higher than 1.0 when the condensation temperature is below 10 °C. If the ORC is designed for single-functional system in cold areas, single-stage expander will be not preferred because high Mach number may lead to shock waves and low efficiencies [30]. On the other hand, the Mach number at the condensation temperature of 50 °C is only about 0.5 even when evaporation temperature is 140 °C. In this case the design of expander will be much easier.

Table 3.5 Parameter distributions in an expectable ORC-based CHP system

	State point	Pump outlet	Evaporator outlet	Expander inlet	Expander outlet	Condenser outlet
$T_h = 110$ °C	t (°C)	55.9	108.8	108.3	75.5	55.0
$\eta_{en,ORC} = 6.1$ %	p (kPa)	976	951	931	272	247
	h (kJ/kg)	257.4	444.3	444.3	429.6	256.3
	s (J/kg/°C)	1,189	1,690	1,691	1,709	1,187

(iii) Aside from the expander, the ORC in the CHP generation has higher exergetic efficiency which means the ORC can reach a higher degree of thermodynamic perfection when operating in the CHP mode.

3.5 Further Discussion

The CHP generation is advantageous regarding the simpler design of the expander and higher thermodynamic perfection of the system. Owing to the low temperature of the heat source, the power efficiency of the ORC for the CHP generation in this test was low. Notably, the data were the results of the preliminary test. The system consisted of inefficient generator, gearbox and pump, which were not supposed to be used in the ORC. And the expander was a preliminary design device. The room for the performance improvement of these components is large. In a commercially applicable ORC, the thermodynamic irreversibility in the flow meter, separator, gearbox, expander, pump, etc. will be avoided or reduced.

An expectable ORC-based CHP system with improved device performance may has parameter distributions in Table 3.5. The working fluid is R123. The expander, the pump and the generator efficiencies are 0.7, 0.5 and 0.85 respectively. The pressure losses in the evaporator, condenser and pipes are taken into consideration. The hot side temperature T_h denotes the maximum temperature in the evaporator. The electricity efficiency of 6.1 % is calculated by Eqs. (3.4) and (3.5). In a similar way of calculation, the efficiency of 7.6, 8.7 and 9.3 % is for T_h = 130, 150 and 170 °C respectively. The efficiency from 6.1 to 9.3 % may not be so high as that of large-scale ORCs with high operation temperature, but will be adequate for small-scale CHP systems for residential buildings. Take Tibet and Qinghai for example. They are two provinces of China which have the richest solar energy resource. There are more than 2 million people in the rural areas without access to electricity [31]. The annual average temperature is about 8–11°C. There is demand on heating more than 180 days/year. And the load is about 70 W per square area of the building. Radiant floor heating and Chinese Kang are the popular technologies in the two areas at present [32]. For a family with building area of 300 m^2, an ORC of output heat of about 20 kW will meet the demand. Water of about 50 °C from the condenser can be served as the source for radiant floor heating and Chinese Kang. An electricity power of 1–2 kW is available, which is enough for television, computer, lighting and video player, etc.

References

1. Qiu GQ, Shao YJ, Li JX, Liu H, Riffat SB (2012) Experimental investigation of a biomass-fired ORC-based micro-CHP for domestic applications. Fuel 96:374–382

2. Li J, Pei G, Li YZ, Ji J (2013) Analysis of a novel gravity driven organic Rankine cycle for small-scale cogeneration applications. Appl Energy 108:34–44
3. The 2nd international seminar on ORC power systems. http://www.asme-orc2013.nl/content/presentations. Accessed 7 Oct 2013
4. Cho SY, Cho CH, Ahn KY, Lee YD (2014) A study of the optimal operating conditions in the organic Rankine cycle using a turbo-expander for fluctuations of the available thermal energy. Energy 64:900–911
5. Walnum HT, Rohde D, Ladam Y (2012) Off-design analysis of ORC and CO_2 power production cycles for low-temperature surplus heat recovery. Int J Low Carbon Technol 0:1–8
6. Wang JF, Yan ZQ, Zhao P, Dai YP (2014) Off-design performance analysis of a solar-powered organic Rankine cycle. Energy Convers Manage 80:150–157
7. Manente G, Toffolo A, Lazzaretto A, Paci M (2013) An organic Rankine cycle off-design model for the search of the optimal control strategy. Energy 58:97–106
8. Twomey B, Jacobs PA, Gurgenci H (2013) Dynamic performance estimation of small-scale solar cogeneration with an organic Rankine cycle using a scroll expander. Appl Therm Eng 51:1307–1316
9. Lecompte S, Huisseune H, van den Broek M, De Schampheleire S, Paepe De M (2013) Part load based thermo-economic optimization of the organic Rankine cycle (ORC) applied to a combined heat and power (CHP) system. Appl Energy 111:871–881
10. Quoilin S, Lemort V, Lebrun J (2010) Experimental study and modeling of an organic Rankine cycle using scroll expander. Appl Energy 87:1260–1268
11. Ibarra M, Rovira A, Alarcón-Padilla D, Blanco J (2014) Performance of a 5 kWe organic Rankine cycle at part-load operation. Appl Energy 120:147–158
12. Fernández FJ, Prieto MM, Suárez I (2011) Thermodynamic analysis of high-temperature regenerative organic Rankine cycles using siloxanes as working fluids. Energy 36:5239–5249
13. Drescher U, Brüggemann D (2007) Fluid selection for the organic Rankine cycle in biomass power and heat plants. Appl Therm Eng 27:223–228
14. Angelino G, Invernizzi C, Molteni G (1997) The potential role of organic bottoming Rankine cycles in steam power stations. Proc Inst Mech Eng Part A J Power Energy 213:75–81
15. Colonna P, Harinck J (2008) Real-gas effects in organic Rankine cycle turbine nozzles. J Propul Power 24:282–294
16. Fiaschi D, Manfrida G, Maraschiello F (2012) Thermo-fluid dynamics preliminary design of turbo-expanders for ORC cycles. Appl Energy 97:601–608
17. Li J, Pei G, Li YZ, Wang DY, Ji J (2013) Examination of the expander leaving loss in variable organic Rankine cycle operation. Energy Convers Manage 65:66–74
18. Zheng N, Zhao L, Wang XD, Tan YT (2013) Experimental verification of a rolling-piston expander that applied for low-temperature organic Rankine cycle. Appl Energy 112:1265–1274
19. Lemort V, Quoilin S, Cuevas C, Lebrun J (2009) Testing and modeling a scroll expander integrated into an organic Rankine cycle. Appl Therm Eng 29:3094–3102
20. Bracco R, Clemente S, Micheli D, Reini M (2013) Experimental tests and modelization of a domestic-scale ORC (organic Rankine cycle). Energy 58:107–116
21. Avadhanulal VK, Lin CS (2014) Empirical models for a screw expander based on experimental data from organic Rankine cycle system testing. J Eng Gas Turbines Power 136:062601–062608
22. Erhart TG, Eicker U, Infield D (2013) Influence of condenser conditions on organic Rankine cycle load characteristics. J Eng Gas Turbines Power 135:042301–042309
23. Bejan A (2006) Advanced engineering thermodynamics, 3rd edn. Wiley, Hoboken
24. Li J, Pei G, Li YZ, Wang DY, Ji J (2012) Energetic and exergetic investigation of an organic Rankine cycle at different heat source temperatures. Energy 38:85–95
25. Liu H, Qiu GQ, Shao YJ, Daminabo F, Riffat SB (2010) Preliminary experimental investigations of a biomass-fired micro-scale CHP with organic Rankine cycle. Int J Low Carbon Technol 5:81–87

26. Riffat SB, Zhao XD (2004a) A novel hybrid heat pipe solar collector/CHP system—Part I: System design and construction. Renew Energy 29:2217–2233
27. Riffat SB, Zhao XD (2004b) A novel hybrid heat-pipe solar collector/CHP system—Part II: Theoretical and experimental investigations. Renew Energy 29:1965–1990
28. Wang XD, Zhao L, Wang JL, Zhang WZ, Zhao XZ, Wu W (2010) Performance evaluation of a low-temperature solar Rankine cycle system utilizing R245fa. Sol Energy 84:353–364
29. James AM, Jon RJ, Jiming C, Douglas KP, Richard NC (2009) Experimental testing of gerotor and scroll expanders used in, and energetic and exergetic modeling of, an organic Rankine cycle. J Energy Res Technol 131:012201–012208
30. Gronman A (2010) Numerical modeling of small supersonic axial flow turbines. Dissertation thesis, Lappeenranta University of Technology. ISBN 978-952-214-953-4
31. http://data.worldbank.org/indicator/EG.ELC.ACCS.ZS. Accessed 20 Nov 2013
32. Zhi Z, Yuguo L, Bin C, Jiye G (2009) Chinese kang as a domestic heating system in rural northern China—a review. Energy Build 41:111–119

Chapter 4
Examination of Key Issues in Designing the ORC Condensation Temperature

The design of the condensation temperature is crucial to the annual power conversion of the ORC. For an ORC that only generates power, the operating condensation temperature fluctuates greatly in many areas through the year due to the variation of environment temperature. For an ORC in the CHP application, the fluctuation of the condensation temperature is also unavoidable in view of the seasonality of consumers' demand on heat. In the no demand periods the system shall work under lower condensation temperature for more efficient power generation. Off-design operation of the ORC will be executed, accompanied with a degraded performance of the components especially the expander. The design of the condensation temperature then influences the efficiency in both CHP generation and solar power generation (SPG). If the condensation temperature is designed at a high value or simply based on the CHP generation, the power conversion will suffer from low expander efficiency at low operating condensation temperature. An optimum design condensation temperature shall be a compromise among the power outputs over a wide range of operating condensation temperature. This chapter aims to determine the optimum design condensation temperature for the ORC. The key issues in the design i.e. the expander characteristics, ratio of operation times in CHP and SPG modes, evaporation temperature, and annual environment temperature are examined.

4.1 Background

As demonstrated in Chap. 3, domestic CHP generation is one preferred application of the ORC. The domestic ORC in the power range from several kW to a few hundreds of kW is attracting increasing attention [1–5]. It avoids large scale collection and storage of renewable energy resources such as solar energy and biomass energy, and generates both heat and power near the point of usage. It is particularly attractive for high latitude areas and remote regions without the access to electricity.

One challenge along with the domestic ORC-CHP system is the consumer's demands on heat fluctuate with the season. Unlike the large scale steam Rankine cycle based CHP system, the domestic ORC-CHP system is characterized by highly

J. Li, *Structural Optimization and Experimental Investigation of the Organic Rankine Cycle for Solar Thermal Power Generation*, Springer Theses,
DOI 10.1007/978-3-662-45623-1_4

variable operation through the year [6, 7]. The heat source for the system is generally unsteady. Solar radiation fluctuates with the time of day. The heating value of biomass varies considerably between species, moisture content, etc. The heat sink thermal performance also changes with the weather condition. All these factors have made the variable operation of the ORC-CHP system an important research field recently [8–14]. The variable operation will be executed, which complicates the design of the system.

Particularly, the domestic ORC-CHP system will inevitably undertake a wide range of condensation temperature regarding the seasonality of consumers' demand on heat. Take China for example, the longest heating in a year takes place in Hulunbuir Genhe city with a duration of about 8 months, while the heating period is generally less than 6 months for other cities. In the non-heating period, the system shall work under lower condensation temperature for the sake of larger temperature difference driving the ORC and thus more power output. Table 4.1 presents some information on the system power efficiency under the condensation temperature of 20 and 60 °C respectively. The evaporation temperature ranges from 100 to 140 °C. For every evaporation temperature and working fluid, the relative increment of the power efficiency at 20 °C by that at 60 °C is larger than 48.2 %. The increment gets more significant at lower evaporation temperature.

The design of the condensation temperature emerges as a critical issue for the domestic ORC-CHP system. First, it influences the performance of the expander, pump, heat exchanger, etc. in the practical operation. The operating condensation temperature is strongly correlated with the expansion ratio of the expander, the pressurization ratio of the pump and the temperature difference in the heat exchangers. Once a device in the ORC is constructed according to the design condensation temperature, its behavior at other condensation temperatures will obey the performance maps [15]. The expander or the pump at the design condition is supposed to have a maximum efficiency, which is normally restricted by the

Table 4.1 The ORC power efficiency under different conditions of the evaporation temperature, condensation temperature and working fluid, unit: %

Evaporation temperature (°C)	Condensation temperature (°C)	Working fluid				
		R123	Pentane	R141b	Butane	R245fa
100	60	5.34	5.30	5.53	4.83	4.99
	20	10.25	10.10	10.59	9.51	9.66
110	60	6.28	6.24	6.55	5.58	5.78
	20	10.96	10.79	11.37	10.09	10.26
120	60	7.10	7.04	7.45	6.17	6.44
	20	11.58	11.40	12.08	10.54	10.74
130	60	7.80	7.72	8.24	6.62	6.93
	20	12.12	11.93	12.70	10.87	11.12
140	60	8.38	8.30	8.93	6.84	7.23
	20	12.57	12.38	13.24	11.05	11.34

Note The efficiencies of the expander, pump and generator are 0.7, 0.5 and 0.85 respectively

technical level. The efficiency decreases when the pressure ratio deviates from the design [16, 17], which can result from the variation of the condensation temperature. Second, for different meteorological conditions, the condensation temperature at design will be different. The design needs to be consistent with the local climate. In areas of short heating season the condensation temperature at design shall be lower than that for the CHP generation. Third, the operating condensation temperature in the SPG mode is lower than that in the CHP mode. If the condensation temperature of the system is designed based on the SPG mode, the ORC will experience off-design operation in the CHP mode and lead to degradation of the power efficiency. The opposite is also true. If the condensation temperature is designed based on the CHP mode, the ORC will suffer from degraded efficiency in the SPG mode. The optimum design condensation temperature is the compromise between the outputs in the SPG and CHP modes.

The design of the ORC condensation temperature is important to the system on the solor purpose of power generation as well. It is common in many areas for an ORC to experience the environment temperature from 0 to 37 °C through a year. The ORC is favorably adopted in low temperature applications, and the variation of the environment temperature will significantly affect the system power efficiency. The annual power output can be deemed as the integral of a function with respect to the operating condensation temperature within a certain interval. And the function changes with the design condensation temperature.

A careful design of the condensation temperature is essential for both domestic ORC-CHP and SPG systems. And the first step is to clarify the performance characteristics of the devices. Among the components in the ORC, the expander plays a key role in the power conversion.

4.2 Introduction to Performance Characteristics of the Expander

The operating condensation temperature influences the pressure ratio of the expander. The efficiency variation with the pressure ratio has been investigated in some works for volumetric expanders, as shown in Fig. 4.1. Avadhanula and Lin built two empirical models for a screw expander based on experimental data [18]. The screw expander worked over a pressure ratio range from 2.70 to 6.54. The expander specific power output in Model I and Model II was expressed by Eqs. (4.1) and (4.2) respectively,

$$w_{SE} = \frac{n_c P_4 v_4}{n_c - 1}[1 - \frac{r_v}{r_p}] \tag{4.1}$$

$$w_{SE} = h_o(L * w_{s,o}^4 + M * w_{s,o}^3 + N * w_{s,o}^2 + O * w_{s,o} + P) \tag{4.2}$$

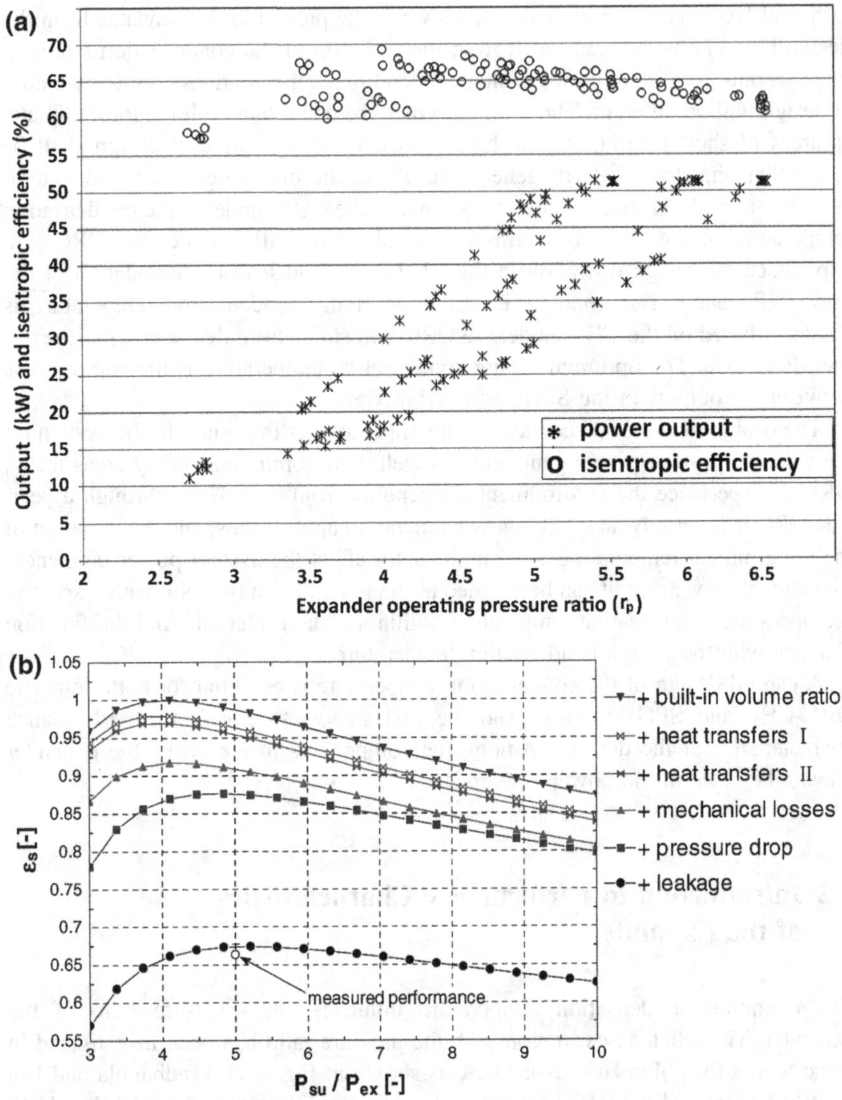

Fig. 4.1 Variations of the isentropic efficiency with the pressure ratio for volumetric expanders: test results by **a** Avadhanula and Lin Reprinted from Ref. [18]; **b** Lemort et al. Reprinted from Ref. [19], Copyright 2009, with permission from Elsevier; **c** Bracco et al. Reprinted from Ref. [1], Copyright 2013, with permission from Elsevier; **d** Zhu et al. Reprinted from Ref. [20], Copyright 2014, with permission from Elsevier

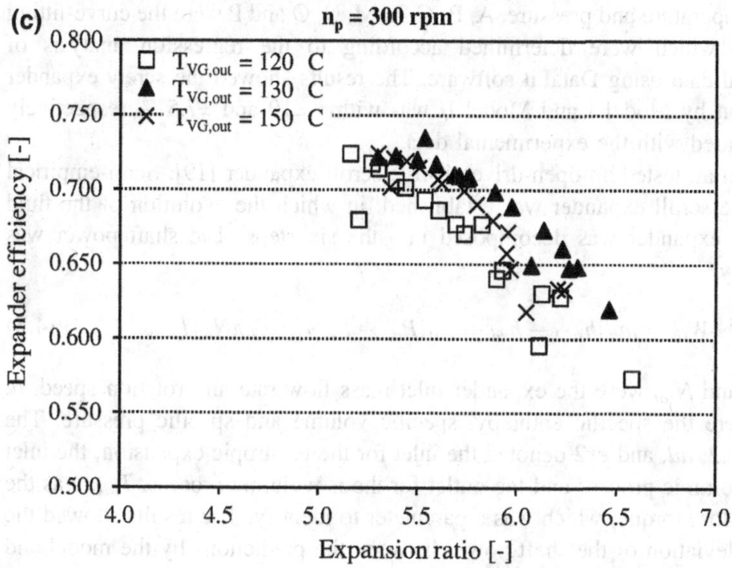

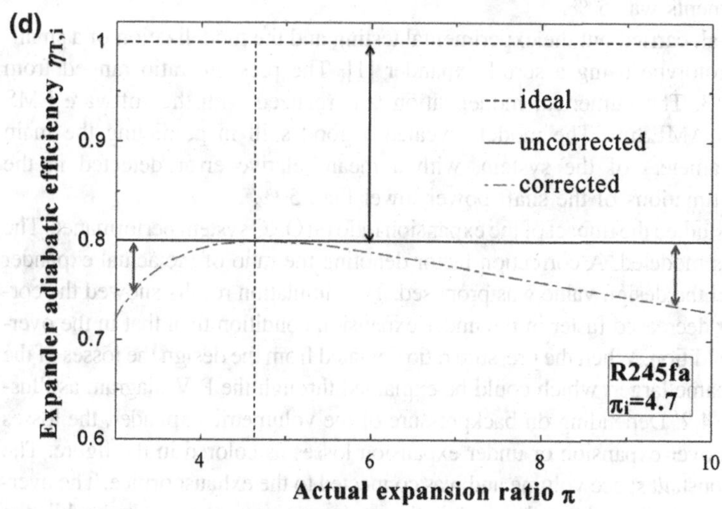

Fig. 4.1 (continued)

$$n_c = A\frac{\ln r_p}{\ln r_v} + B(\frac{\ln r_p}{\ln r_v})^2 + C(\frac{\ln r_p}{\ln r_v})^3 \tag{4.3}$$

$$w_{s,o} = \frac{w_s}{h_o} \tag{4.4}$$

where P_4, v_4, r_v, r_p and w_s were the expander inlet pressure, inlet specific volume, pressure ratio, volume ratio, and isentropic power output. h_o was the enthalpy at standard temperature and pressure. A, B, C, L, M, N, O and P were the curve-fitting coefficients, which were determined according to the regression analysis of experimental data using DataFit software. The results showed the screw expander power output by Model I and Model II was within ±10 and ±7.5 % respectively when compared with the experimental data.

Lemort et al. tested an open-drive oil-free scroll expander [19]. Semi-empirical model of the scroll expander was established, in which the evolution of the fluid through the expander was decomposed into the six steps. The shaft power was expressed by

$$W_{sh} = m_{in}\left[h_{su,2} - h_{ad} + v_{ad}\left(P_{ad} - P_{ex,2}\right)\right] - 2\pi N_{rot} T_{loss} \tag{4.5}$$

where m_{in} and N_{rot} were the expander inlet mass flow rate and rotation speed. h, v and P were the specific enthalpy, specific volume and specific pressure. The subscripts $su2$, ad, and $ex2$ denoted the inlet for the isentropic expansion, the inlet for the isovolumic process and the outlet for the isovolumic process. T_{loss} was the mechanical loss torque, which was a parameter to identify. The results showed the maximum deviation of the shaft power between the predictions by the model and the measurements was 5 %.

Bracco et al. carried out the experimental testing and the modelization of a small-size ORC prototype using a scroll expander [1]. The pressure ratio ranged from about 5 to 6.3. The numerical modelization was realized with the software LMS Imagine Lab AMESim. The model revealed a good skill in predicting the main working parameters of the system, with a mean relative error detected in the numerical estimations of the shaft power lower than 5 %.

Zhu et al. studied the impact of the expansion ratio on ORC system performance. The expander was modeled. A correction factor denoting the ratio of the actual expander efficiency and the design value was proposed. The simulation results showed the correction factor decreased faster in the under-expansion condition than that in the over-expansion condition. When the pressure ratio deviated from the design the losses of the expander became larger, which could be explained through the P-V diagram as illustrated in Fig. 4.2. Depending on backpressure of the volumetric expander, the losses stood for the over-expansion or under-expansion losses as colored in the figure. The tooth had a constant space volume and was connected to the exhaust orifice. The over-expansion losses indicated that the work in the green area was not converted, while the under-expansion losses meant the work was negative [20].

There is also some information on the efficiency variation with the pressure ratio for turbo expanders driven by air, as shown in Fig. 4.3. Vlasic et al. presented the design and performance of a high work research turbine (HWRT) [21]. The HWRT design was aggressive in terms of pressure ratio, stage loading, rotor trailing edge blockage and flow area speed square. The performance of the turbine was evaluated through measurements of reaction, rotor exit conditions and efficiency, with and without airfoil cooling. The rig inlet pressure and temperature were kept constant and set at 275 kPa and 191 °C, respectively. A sharp drop-off in efficiency was observed. The rapid drop in turbine efficiency at the design speed began at a rotor

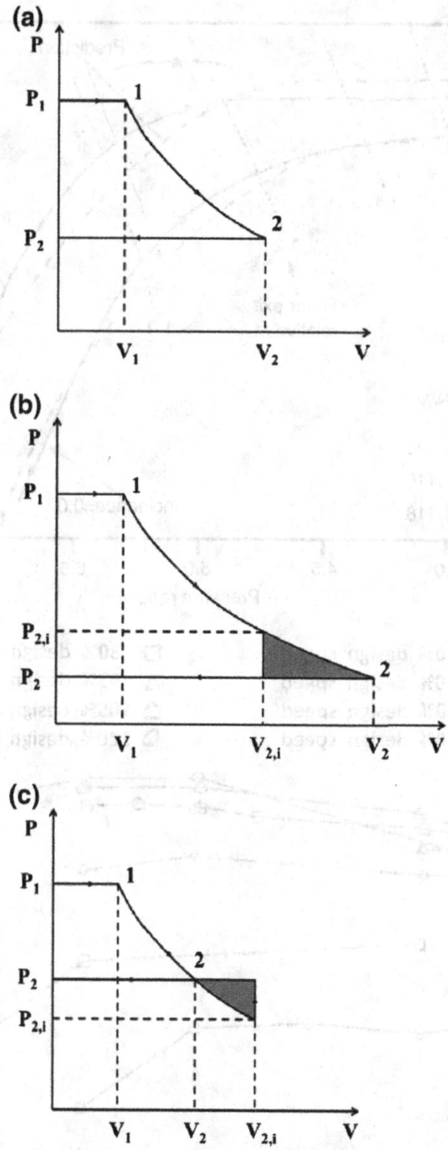

Fig. 4.2 Losses of a volumetric expander when the actual pressure ratio differs from the design: P_2—the actual backpressure, $P_{2,i}$—the design. Reprinted from Ref. [20], Copyright 2014, with permission from Elsevier. **a** $P_2 = P_{2,i}$, normal process, **b** $P_2 < P_{2,i}$, over-expansion, **c** $P_2 > P_{2,i}$, under-expansion

exit relative Mach number of about 1.15 with a design pressure ratio of 4.5. The characteristics of the single-stage turbine stage, and in particular the rotor trailing edge blockage and stage reaction, were the reason for this rapid decrease.

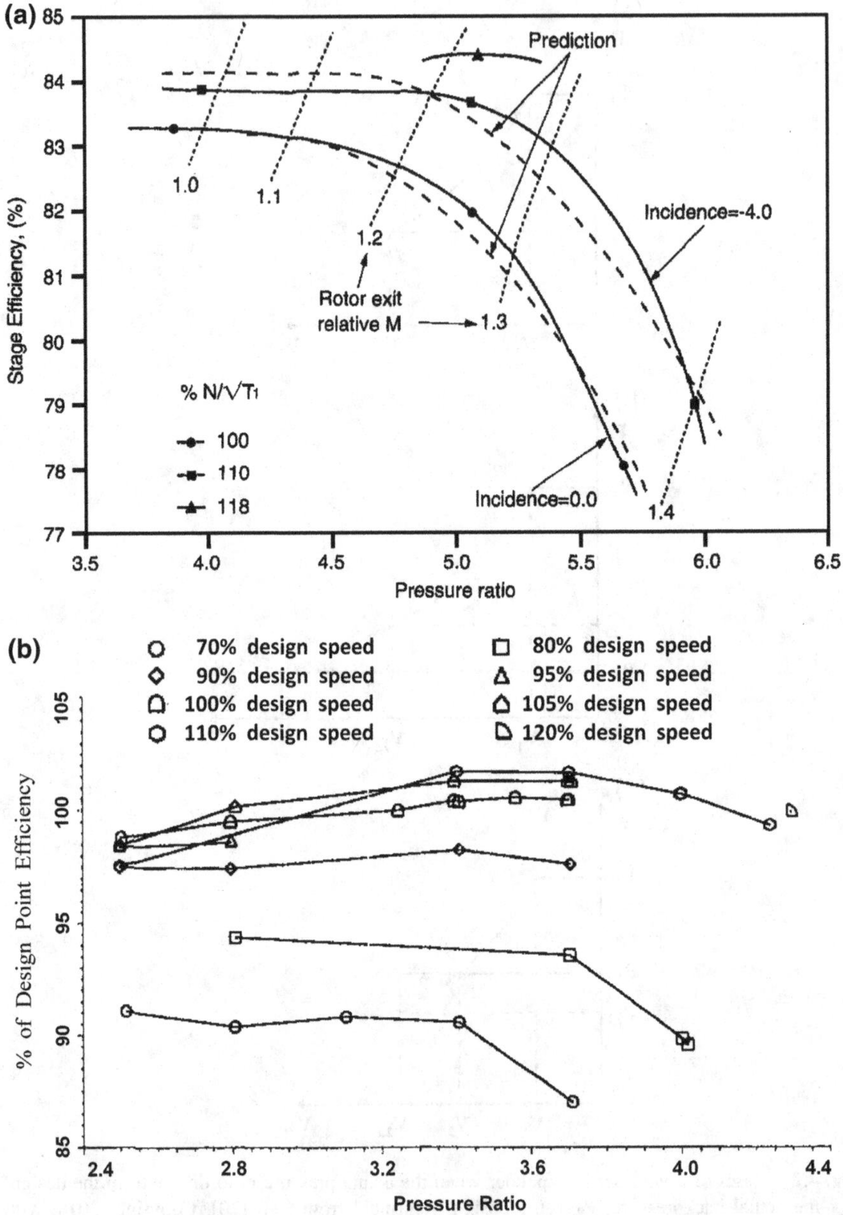

Fig. 4.3 Isentropic efficiency variations with the pressure ratio for turbo expanders: test results by **a** Vlasic et al. Reprinted from Ref. [21], Copyright 1996, with permission from ASME; **b** Woinowsky et al. Reprinted from Ref. [22], Copyright 1999, with permission from ASME

Woinowsky-Krieger et al. presented the aerodynamic design and cold flow rig testing of a single-stage compressor turbine with transonic aerofoils and a low-Reynolds-number blade [22]. This kind of turbine had the advantage of light weight, low initial cost and easier maintenance. The turbine was fed by a plenum of air, and the inlet pressure was set to reproduce the engine equivalent sea-level or altitude stage Reynolds number. The outlet pressure of the turbine was adjusted by two compressors, which could pull the exhaust pressure down to 13.8 kPa, allowing for the range of stage pressure ratios and Reynólds numbers necessary for mapping the turbine characteristics. The effect of the pressure ratio on the turbine efficiency at off-design conditions was investigated. Seven efficiency-pressure ratio curves were presented, with corresponding rotation speed of 70, 80, 90, 95, 100, 105, 110 % of the design speed. The results showed the relative deviation of the efficiency was within 2 % when the deviation of the pressure ratio was within 20 % of the design value. The curves were quite flat in the interval of pressure ratio from 2.5 to 3.5. But drastic declines were formed when the pressure ratio exceeded 4.0.

Above all, the variation of the expander efficiency with the pressure ratio is evident and significant. The curves open downward. For each curve, there will be a pressure ratio at which the expander efficiency reaches the maximum. The shapes of curves for the volumetric expanders are distinguishable from those for the turbo expanders. For the former, the variation of efficiency at higher pressure ratio seems smoother than that at lower pressure ratio as shown in Fig. 4.1. For the latter, the variation is very slight when the pressure ratio is lower than the design value, making the left part of the curve much smoother than the right part.

This chapter is focused on the design of the ORC condensation temperature with respect to the performance characteristics of the expander. The expander model is established. The design methodology of the condensation temperature in various operating environments such as the evaporation temperature, operation mode and annual environment temperature, is discussed. The work is a attempt to realize efficient small scale ORC systems.

4.3 Modeling of the Expander

The evaporation temperature of the ORC and the rotation speed may also affect the efficiency of the expander. However, the evaporation temperature is restricted by the nature of heat source. The design of the condensation temperature is generally independent on the design of the evaporation temperature and the expander rotation speed. To concentrate on the key terms of the condensation temperature designing, this chapter makes the assumption of constant ORC evaporation temperature and expander rotation speed. The two parameters will be invariable during the ORC operation, which comes true when a heat storage unit and electric equipment of a certain frequency are used.

On the given conditions of the evaporation temperature and rotation speed, the relationship between the expander efficiency (ε_t) and the pressure ratio (π_t) differs from one expander to another as shown in Figs. 4.1 and 4.2. The mathematical function $\varepsilon_t = f(\pi_t)$ for each of the curves is complicated. At present a universal, well-demonstrated model dealing with the expander efficiency over a wide range of pressure ratio is unavailable. And the aforementioned works in Sect. 4.2 deal with either the volumetric expanders or the air-driven turbo expanders. Information of the turbo expander efficiency varying with the pressure ratio for ORC application is lacked. To design the ORC condensation temperature, the characteristics for a 3.5 kW turbo expander using the working fluid of R123 will be mapped and modeled. This turbo expander has been designed and manufactured specially for applying in the ORC. This kind of expander has many advantages such as a compact structure with good manufacturability, high ratio of power to volume, a single-stage expansion rate of large enthalpy drop, and high efficiency. Specifics of the expander and the ORC test rig can be achieved in some works by the authors [23, 24].

Figure 4.4 shows the efficiency variation of the turbo expander when the rotation speed is about 36,000 rpm. The inlet temperature and pressure are about 100 °C and 0.5 MPa respectively. The polynomial fit for the experiment data is also depicted in the figure. The polynomial function is expressed by Eq. (4.6). According to this equation, the maximum expander efficiency is 0.5966 with a pressure ratio of 4.7937 [25].

$$\varepsilon_t = 0.5966 - 0.1025\left(\frac{\pi_t}{4.7937} - 1\right)^2 \tag{4.6}$$

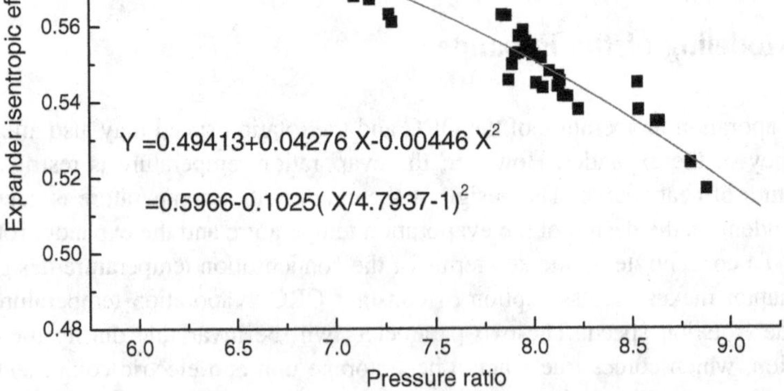

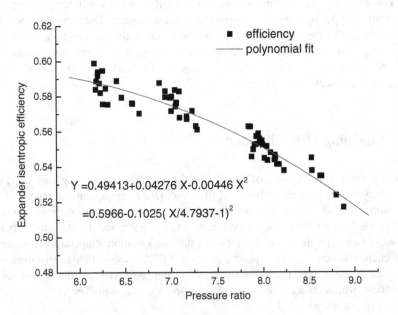

Fig. 4.4 The turbo expander isentropic efficiency variation under a rotation speed around 36,000 rpm. Reprinted from Ref. [25], Copyright 2014, with permission from Elsevier

Equation (4.6) is a rudimental model for the expander. The extensive form is adopted in this chapter, which is expressed by Eq. (4.7). $\pi_{t,0}$ is the design pressure ratio. a and b are the coefficients.

$$\varepsilon_t = a - b \cdot \left(\frac{\pi_t}{\pi_{t,0}} - 1\right)^2 \tag{4.7}$$

According to Eq. (4.7), the expander efficiency at a given pressure ratio is affected by the design pressure ratio. The expander efficiency decreases as the deviation of pressure ratio from the design gets larger. Notably, the original function revealing the substantial relationship between the expander efficiency and pressure ratio is complicated, but it can be expressed as an infinite sum of terms by Taylor series expansion $f(\pi_t) = \sum_{n=0}^{+\infty} \frac{f^{(n)}(x_0)}{n!}(\pi_t - x_0)^n$. And Eq. (4.7) can be deemed as a secondary approximation of $f(\pi_t)$. This equation might not be applicable for other expanders, other operation conditions, or extreme deviation of the pressure ratio. However, this chapter draws interest in π_t from about 2.8 (the ratio of the saturation pressure of R123 at 100 °C and that at 60 °C) to 30 (the ratio of the saturation pressure of R123 at 160 °C and that at 20 °C). In this interval, a secondary approximation can maintain adequate accuracy. Pursuant to this approximation, the variation of the expander efficiency with the pressure ratio is shown in Fig. 4.5. The design pressure ratio is 10. The decrement of the expander efficiency is relatively smooth when the pressure ratio is lower than the design value, which is consistent with the experiment results by

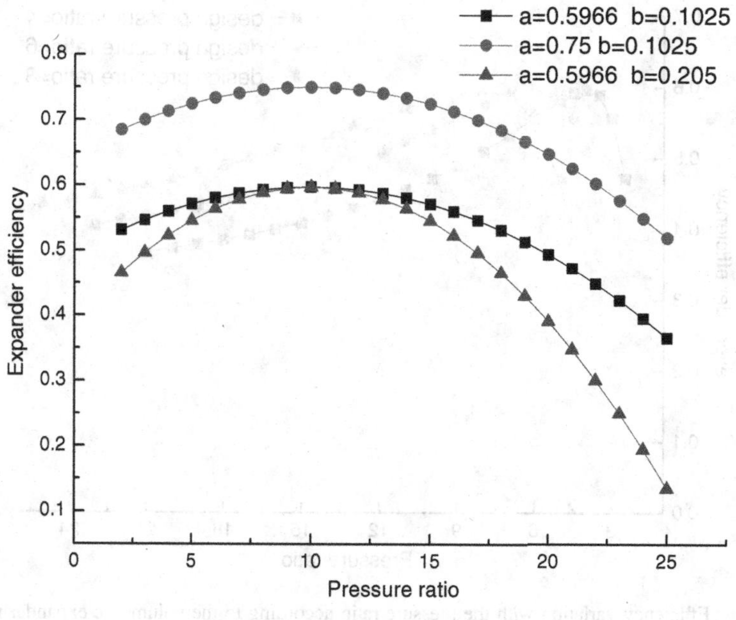

Fig. 4.5 Efficiency variation with the pressure ratio according to the turbo expander model

Vlasic and Woinowsky et al. It implies no extra work of the expander is possible when the pressure ratio exceeds a certain value, which can be called as the limit-load pressure ratio. The locus of expander limit-load point is an important constraint for off-design analysis. In concept, given the inlet temperature and pressure, there is a maximum amount of power that can be produced by an expander and a corresponding maximum pressure ratio that realizes efficient expansion. This condition is referred to as limit load of the expander. For single stage turbo expander, limit load occurs after the rotor has choked at its throat, and further expansion causes the axial Mach number across its exit to become unity. Beyond this point, the downstream of the rotor has no influence on the flow through the rotor, and the increment in pressure ratio no longer enlarges the turbo expander output power [26].

Aside from the turbo expander model, by exchanging the positions of $\pi_{t,0}$ and π_t, a preliminary model for the volumetric expander can be obtained as expressed by Eq. (4.8).

$$\varepsilon_t = a - b \cdot \left(\frac{\pi_{t,0}}{\pi_t} - 1\right)^2 \tag{4.8}$$

The expander efficiency variation with the pressure ratio according to Eq. (4.8) is displayed in Fig. 4.6. Sharp decreases at low pressure ratios show up, which is in agreement with the results in the previous works [19, 20].

The general variations of the efficiency with the pressure ratio are exhibited by Eqs. (4.7) and (4.8) for the turbo and volumetric expanders. It shall be pointed out

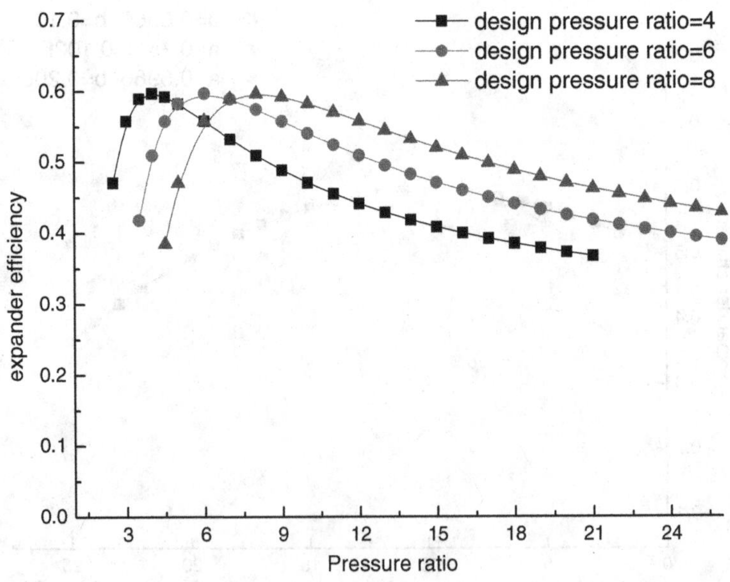

Fig. 4.6 Efficiency variation with the pressure ratio according to the volumetric expander model of $a = 0.5966$ and $b = 0.35$

that modeling of the expander is a precondition of the condensation temperature design. Equations (4.7) and (4.8) just serve as examples. In the following sections the key issues in design of the ORC condensation temperature are examined. The conclusions of the examination are expected be applicable for other relationships between the expander efficiency and the pressure ratio.

4.4 Variation of the Efficiency in the ORC Power Conversion with the Condensation Temperature in Regard to the Expander Characteristics

The design of the ORC condensation temperature is subjected to the expander characteristics. Without the variation of the expander efficiency, the design would become unnecessary because the ORC performances would be then just influenced by the practical operating temperatures. To make out the role of the expander characteristics in designing the condensation temperature, the ORC power efficiency is investigated on different conditions of *a* and *b* in the expander models. Figures 4.7, 4.8, 4.9, 4.10 and 4.11 show the variations of the turbo expander efficiency and ORC power efficiency when the design condensation temperature is 20, 30, 40, 50 and

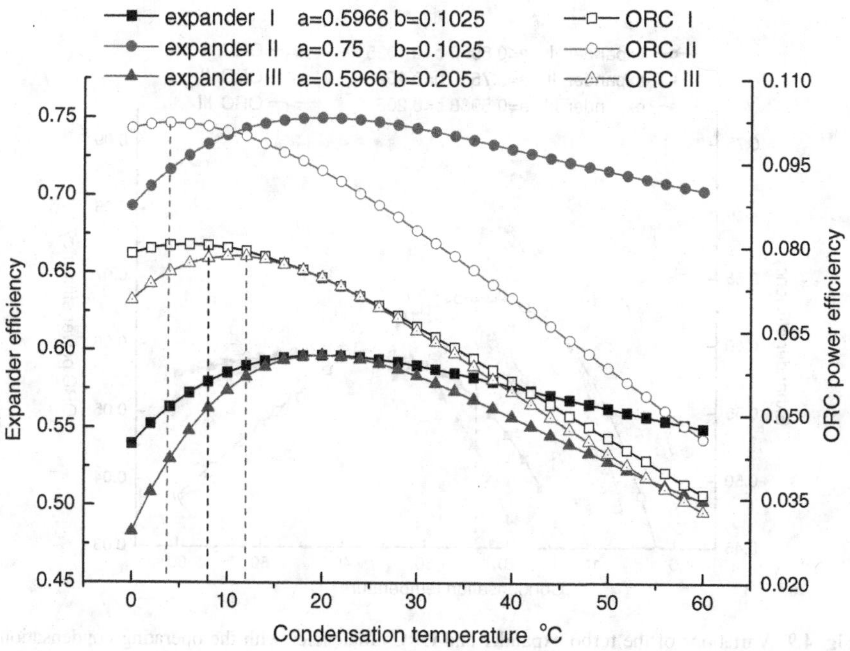

Fig. 4.7 Variations of the turbo expander and ORC efficiencies with the operating condensation temperature for a design value of 20 °C

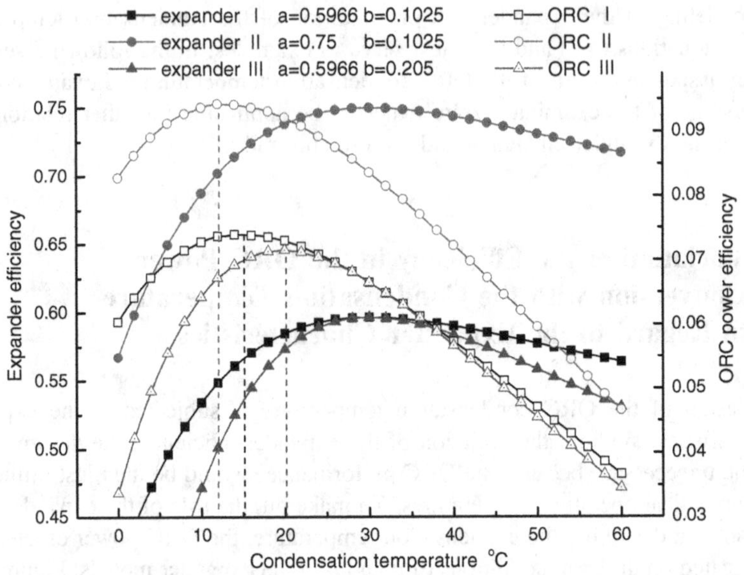

Fig. 4.8 Variations of the turbo expander and ORC efficiencies with the operating condensation temperature for a design value of 30 °C

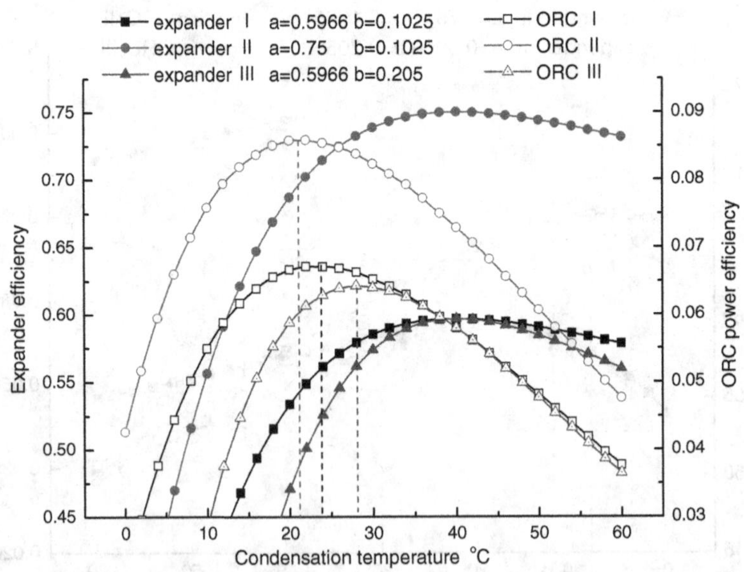

Fig. 4.9 Variations of the turbo expander and ORC efficiencies with the operating condensation temperature for a design value of 40 °C

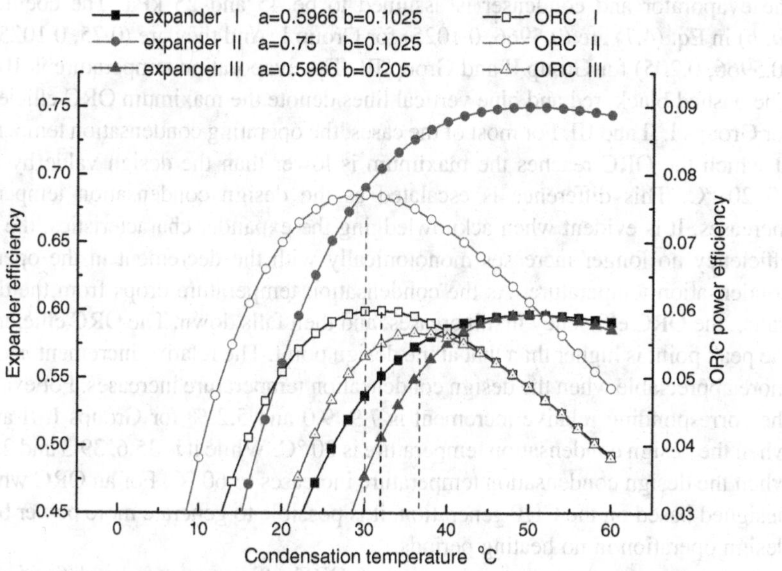

Fig. 4.10 Variations of the turbo expander and ORC efficiencies with the operating condensation temperature for a design value of 50 °C

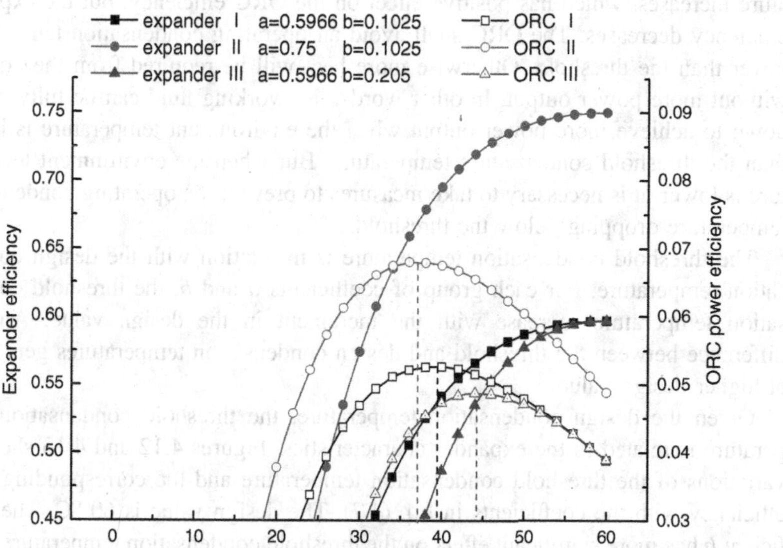

Fig. 4.11 Variations of the turbo expander and ORC efficiencies with the operating condensation temperature for a design value of 60 °C

60 °C. The efficiency for the pump and generator is 0.5 and 0.85. The pressure loss in the evaporator and condenser is assumed to be 45 and 25 kPa. The coefficients (*a*, *b*) in Eq. (4.7) are (0.5966, 0.1025) for Group I. And they are (0.75, 0.1025) and (0.5966, 0.205) for Group II and Group III. The evaporation temperature is 100 °C. The dashed black, red and blue vertical lines denote the maximum ORC efficiencies for Groups I, II and III. For most of the cases, the operating condensation temperature at which the ORC reaches the maximum is lower than the design value by about 10–20 °C. This difference is escalated as the design condensation temperature increases. It is evident when acknowledging the expander characteristics, the ORC efficiency no longer increases monotonically with the decrement in the operating condensation temperature. As the condensation temperature drops from the design value, the ORC efficiency first increases, and then falls down. The ORC efficiency at the peak point is higher than that at the design point. The relative increment becomes more appreciable when the design condensation temperature increases. For example, the corresponding relative increment is 7.9, 9.0 and 5.2 % for Groups I, II and III when the design condensation temperature is 20 °C. While it is 35.6, 39.3 and 25.5 % when the design condensation temperature increases to 60 °C. For an ORC which is designed based on the CHP generation, it is possible to generate more power by off-design operation in no heating periods.

The temperature at the peak point of ORC efficiency can be defined as the threshold condensation temperature (T_{thres}). The threshold condensation temperature is a trade-off between the expander efficiency and the available temperature difference driving the ORC. As the operating condensation temperature gets lower than the design, the difference between the evaporation and condensation temperature increases, which has positive effect on the ORC efficiency, but the expander efficiency decreases. The ORC shall avoid an operating condensation temperature lower than the threshold. Otherwise more heat will be required from the hot side without more power output. In other words, the working fluid can be fully cooled down to achieve more power output when the environment temperature is higher than the threshold condensation temperature. But when the environment temperature is lower, it is necessary to take measures to prevent the operating condensation temperature dropping below the threshold.

The threshold condensation temperature is in relation with the design condensation temperature. For each group of coefficients *a* and *b*, the threshold condensation temperature increase with the increment in the design value. And the difference between the threshold and design condensation temperatures gets larger at higher design value.

Given the design condensation temperature, the threshold condensation temperature is related to the expander characteristics. Figures 4.12 and 4.13 show the variations of the threshold condensation temperature and the corresponding ORC efficiency with the coefficients in Eq. (4.7). The design value is 60 °C. The coefficient *b* has more significant effect on the threshold condensation temperature when compared with *a*. As *b* decreases, the threshold condensation temperature drops dramatically, accompanied with a remarkable increment in the maximum ORC efficiency. It can be deduced that as *b* approaches to zero the variation of expander

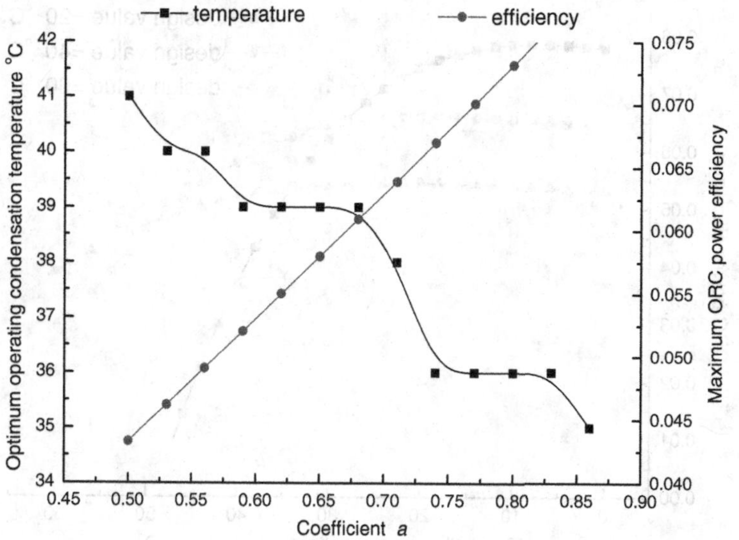

Fig. 4.12 Variations of the optimum operating condensation temperature and maximum ORC efficiency with the coefficient *a* when $b = 0.1025$

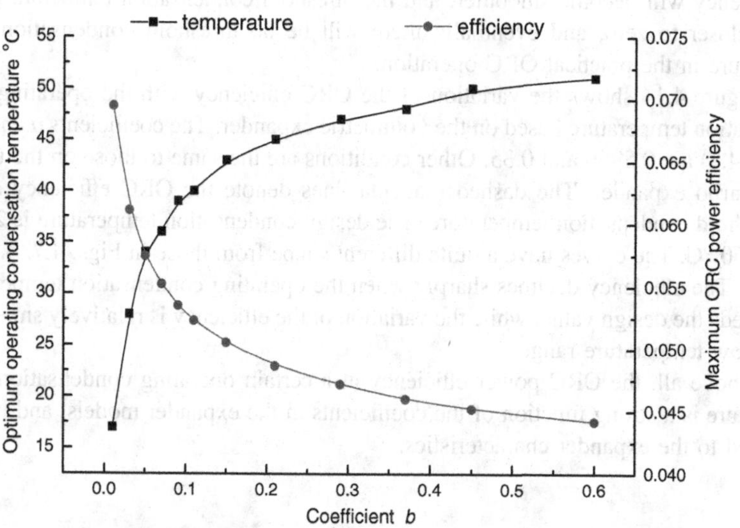

Fig. 4.13 Variations of the optimum operating condensation temperature and maximum ORC efficiency with the coefficient *b* when $a = 0.5966$

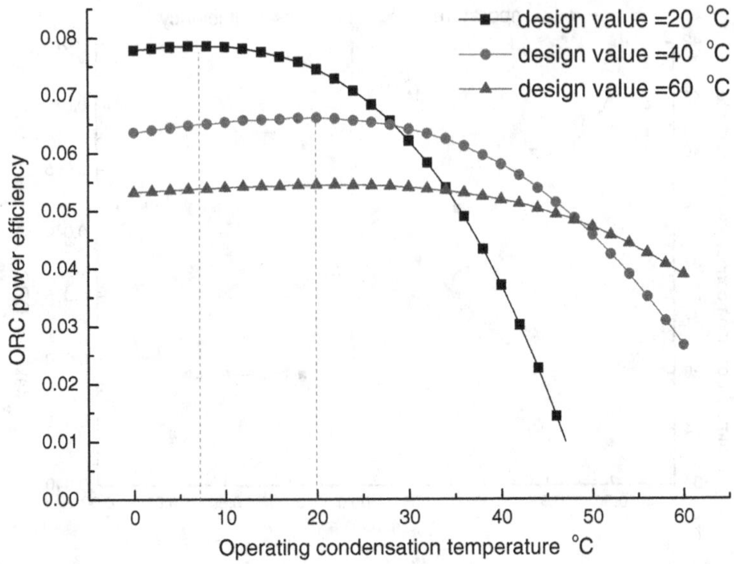

Fig. 4.14 Variation of the ORC efficiency with the operating condensation temperature based on the volumetric expander

efficiency will become smoother, and the threshold condensation temperature will get closer to zero, and eventually there will be no threshold condensation temperature in the practical ORC operation.

Figure 4.14 shows the variation of the ORC efficiency with the operating condensation temperature based on the volumetric expander. The coefficients *a* and *b* in Eq. (4.8) are 0.5966 and 0.35. Other conditions are the same to those on the use of the turbo expander. The dashed magenta lines denote the ORC efficiency at the threshold condensation temperature. The design condensation temperature is 20, 40 and 60 °C. The curves have a quite different shape from those in Figs. 4.7, 4.9 and 4.11. The efficiency declines sharply when the operating condensation temperature exceeds the design value, while the variation of the efficiency is relatively smooth in the low temperature range.

Above all, the ORC power efficiency at a certain operating condensation temperature is a strong function of the coefficients in the expander models, and is thus linked to the expander characteristics.

4.5 Key Issues in the Condensation Temperature Design

A close examination of the expander characteristics is a necessary but not sufficient condition for the optimum design of the ORC condensation temperature. The expander functions upon certain operating environments. And the design of the condensation temperature is influenced comprehensively by the expander characteristics, operation mode, evaporation temperature, local climate, etc. The key issues in the design will be examined in this section. The objective is to realize the most efficient power conversion through the year.

4.5.1 Expander Characteristics

To investigate the influence of the expander characteristics on the design of the condensation temperature, Hefei (E117.27°, N31.86°) is exemplified. The evaporation temperature is 100 °C. As the previous section has pointed out that the ORC operating condensation temperature shall not be lower than the threshold, the information on the threshold condensation temperature under different conditions is necessary for the design work, which is shown in Table 4.2.

The variation of the annual environment temperature in Hefei is depicted in Fig. 4.15. These are the hourly data of the typical meteorological year achieved from Energyplus [27]. The mean environment temperature through the year is 15.5 °C. And the monthly mean environment temperature from January to December is 2.8, 3.9, 8.8, 15.3, 19.8, 23.2, 27.8, 25.6, 22.3, 18.0, 11.3 and 6.2 °C respectively.

The annual power efficiency variation with the design condensation temperature for the turbo and volumetric expanders is shown in Fig. 4.16. During the yearly operation, the ORC has two modes: the SPG in warm seasons and the CHP in cold seasons. The time interval of the SPG is from April 15th to November 15th. The ratio of the SPG operation time to the CHP time is 5159:3601. When the ORC operates in the CHP mode, the condensation temperature is determined by the consumers' demands on heating, bathing, etc. To guarantee human thermal comfort, it is hoped the output heat from the condenser is steady. So the condensation temperature in the CHP mode is assumed to be unchangeable and has a value of 60 °C. On the other hand, when the ORC operates in the SPG mode the available cold side temperature is influenced by the environment temperature ($T_{envir,\ SPG}$), which changes from time to time. In the simulation of the ORC performance in the SPG mode, the operating condensation temperature is the maximum value of the T_{thres} and ($T_{envir,\ SPG}$ +5), that is, max{T_{thres}, $T_{envir,\ SPG}$ +5}. 5 °C represents the temperature difference between the organic fluid and cooling water in the condenser. By this way the power conversion in the SPG mode is optimized. The annual ORC power efficiency is defined as

$$\eta_a = \frac{\eta_{SPG} \cdot t_{SPG} + \eta_{CHP} \cdot t_{CHP}}{t_{SPG} + t_{CHP}} \tag{4.9}$$

where t_{SPG}, t_{CHP} are the operation times for the SPG and the CHP modes.

Table 4.2 The threshold condensation temperature on different conditions

$T_{cond,design}$ (°C)	Turbo expander						Volumetric expander	
	$a = 0.5966$ $b = 0.1025$		$a = 0.75$ $b = 0.1025$		$a = 0.5966$ $b = 0.205$		$a = 0.5966$ $b = 0.35$	
	T_{thres} (°C)	η_{max} (%)	T_{thres} (°C)	η_{max} (%)	T_{thres} (°C)	η_{max} (%)	T_{thres} (°C)	η_{max} (%)
10	<0	/	<0	/	1	8.43	1	8.42
11	<0	/	<0	/	3	8.38	1	8.37
12	<0	/	<0	/	3	8.33	1	8.32
13	0	8.44	<0	/	4	8.26	3	8.27
14	1	8.39	<0	/	5	8.20	3	8.21
15	1	8.33	0	10.56	7	8.15	3	8.16
16	3	8.28	1	10.49	7	8.09	3	8.10
17	3	8.22	1	10.42	7	8.02	5	8.04
18	4	8.16	3	10.36	8	7.96	7	7.98
19	5	8.10	3	10.29	10	7.90	7	7.92
20	7	8.04	4	10.21	11	7.84	7	7.86
21	7	7.99	5	10.13	12	7.77	7	7.80
22	7	7.92	7	10.06	12	7.71	7	7.74
23	8	7.85	7	9.99	14	7.63	11	7.68
24	10	7.79	7	9.91	14	7.57	11	7.62
25	11	7.74	8	9.83	15	7.49	11	7.56
26	11	7.67	8	9.74	17	7.43	11	7.49
27	12	7.61	10	9.67	17	7.37	12	7.43
28	12	7.54	11	9.60	18	7.30	12	7.37
29	14	7.47	12	9.52	18	7.22	12	7.31
30	14	7.40	12	9.43	20	7.16	12	7.24
31	17	7.34	12	9.34	20	7.09	12	7.17
32	17	7.27	14	9.26	22	7.01	14	7.11
33	17	7.21	15	9.17	22	6.94	17	7.05
34	18	7.14	17	9.10	22	6.86	17	6.98
35	20	7.07	17	9.02	24	6.79	17	6.92
36	20	7.01	18	8.93	25	6.72	17	6.86
37	20	6.93	18	8.84	26	6.65	17	6.80
38	22	6.86	20	8.77	27	6.57	17	6.73
39	22	6.79	20	8.68	28	6.50	17	6.67
40	22	6.72	22	8.59	28	6.43	20	6.61
41	25	6.65	22	8.51	28	6.35	20	6.55
42	25	6.58	22	8.41	30	6.27	20	6.49
43	26	6.51	25	8.32	31	6.19	20	6.42

(continued)

Table 4.2 (continued)

$T_{cond,design}$ (°C)	Turbo expander						Volumetric expander	
	$a = 0.5966$ $b = 0.1025$		$a = 0.75$ $b = 0.1025$		$a = 0.5966$ $b = 0.205$		$a = 0.5966$ $b = 0.35$	
	T_{thres} (°C)	η_{max} (%)	T_{thres} (°C)	η_{max} (%)	T_{thres} (°C)	η_{max} (%)	T_{thres} (°C)	η_{max} (%)
44	27	6.43	25	8.24	32	6.12	20	6.36
45	28	6.37	25	8.15	32	6.04	20	6.30
46	28	6.30	26	8.06	34	5.97	20	6.24
47	28	6.22	28	7.98	34	5.90	20	6.18
48	30	6.14	28	7.89	34	5.82	20	6.12
49	30	6.07	28	7.80	36	5.74	20	6.06
50	32	6.00	29	7.70	36	5.66	20	6.00
51	32	5.93	30	7.61	38	5.57	20	5.94
52	34	5.86	32	7.52	39	5.50	20	5.88
53	34	5.79	32	7.43	39	5.43	20	5.83
54	34	5.72	34	7.35	40	5.35	20	5.77
55	36	5.64	34	7.27	41	5.27	20	5.72
56	36	5.57	34	7.18	41	5.19	20	5.66
57	36	5.49	34	7.08	43	5.12	20	5.61
58	38	5.41	36	6.99	43	5.04	20	5.56
59	39	5.34	36	6.90	43	4.96	22	5.50
60	39	5.27	37	6.79	44	4.87	22	5.45

Note T_{thres} is integer, obtained by the discrete algorithm

For each of the curve, there is an optimum design condensation temperature at which the annual ORC power efficiency reaches the maximum. For the turbo expander, the optimum design condensation temperature is about 27 °C, while for the volumetric expander it is about 47 °C. The results indicate that the optimum design condensation temperature can change strongly with the expander characteristics. The maximum efficiency for the black, red, blue and magenta curves is 5.57, 7.03, 5.45 and 5.03 % respectively. And the efficiency at the design condensation temperature of 60 °C is 4.70, 6.0, 4.47 and 4.78 %. The relative increment is about 18.5, 17.2, 21.9, and 5.23 %. It is very clear if the ORC condensation temperature is simply designed based on the CHP operation, the annual power output will be lower than the optimum. And a more efficient ORC can be realized by an accurate design of the condensation temperature.

A comparison between the maximum efficiency and the efficiency under the annual average operating condensation temperature is meaningful. Following the optimum design, the annual average operating condensation temperature is around 40 °C for both the turbo and volumetric expanders-based ORCs. The efficiency at 40 °C for the curves is 5.42, 6.88, 5.26 and 4.91 %, respectively. Therefore, the relative increment of the maximum efficiency by that at 40 °C is only 2.77, 2.18,

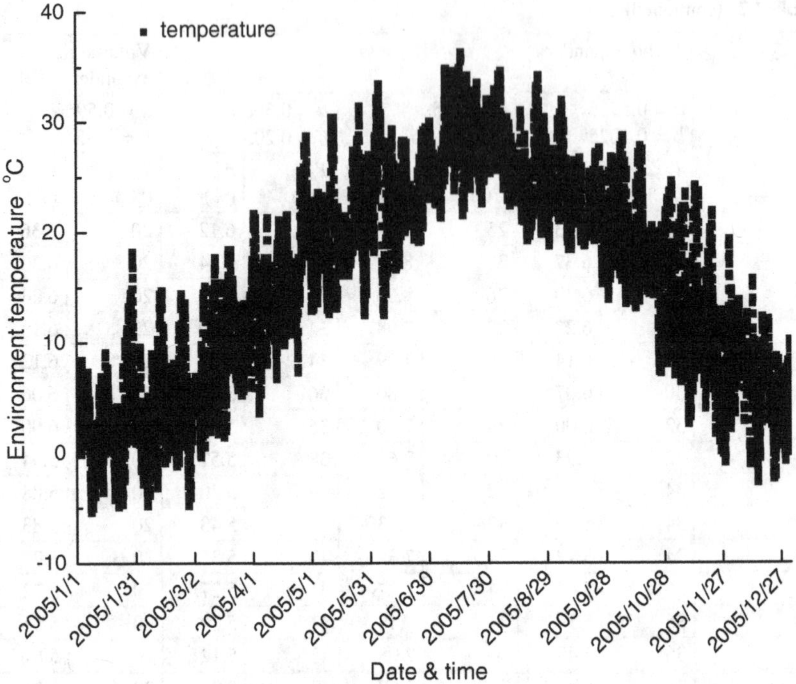

Fig. 4.15 Variation of the annual environment temperature of Hefei in 2005

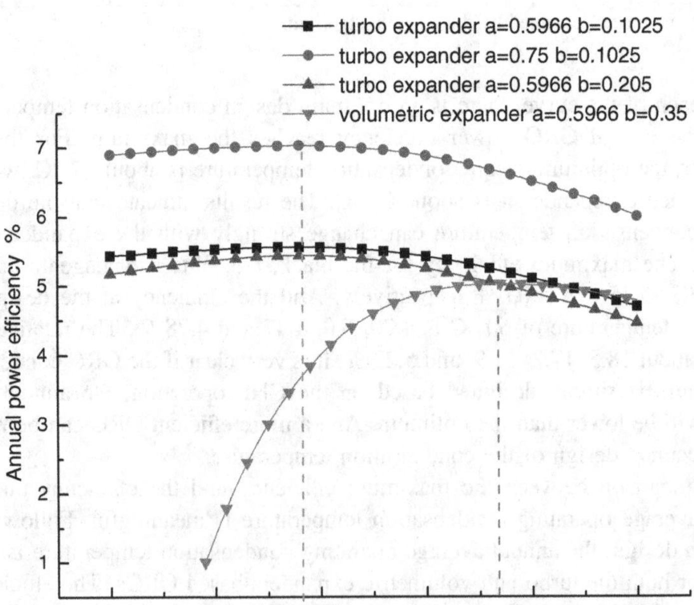

Fig. 4.16 Variation of the annual power efficiency with the design condensation temperature

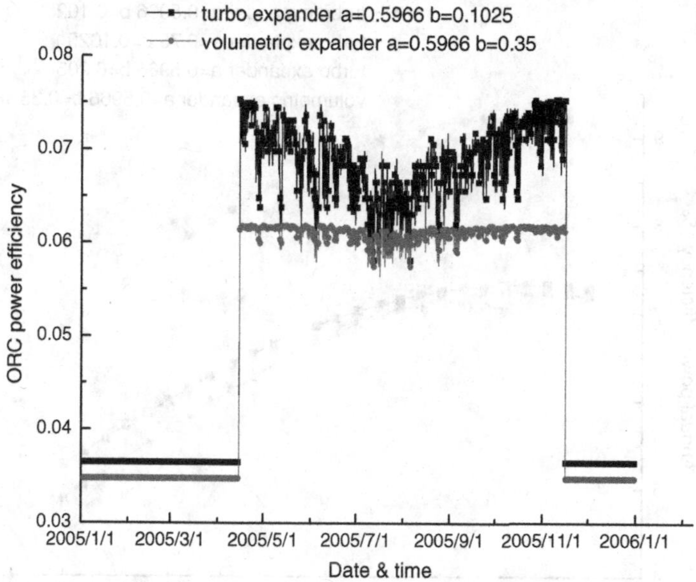

Fig. 4.17 Variation of the hourly power efficiency with time when the design condensation temperature is optimal

3.61 and 2.44 %. The results hint in case there is no access to the expander characteristics, a design condensation temperature of the annual average operating value may be an alternative choice.

The variation of the hourly ORC power efficiency under the optimum design condensation temperature is shown in Fig. 4.17. For both the curves, the power efficiency in the SPG mode is about twice of that in the CHP mode. Though the CHP application is important, heat is not continuously desirable through the year and it is necessary for the ORC to shift its operating mode for the sake of added power output. According to the curves, the ORC driven by the turbo expander has relatively higher efficiency. However, the volumetric expander can offer a more steady output in the SPG mode.

4.5.2 Operation Mode

Figure 4.18 shows the annual power efficiency variation with the design condensation temperature for complete SPG generation. Compared with the curves in Fig. 4.16, the optimum design condensation temperature decreases while the maximum efficiency for each curve increases. The optimum design condensation temperature for the turbo expander is about 16 °C, and it is about 23 °C for the volumetric expander. The annual average operating condensation temperature is about 20 °C. The maximum efficiency for the black, red, blue and magenta curves is

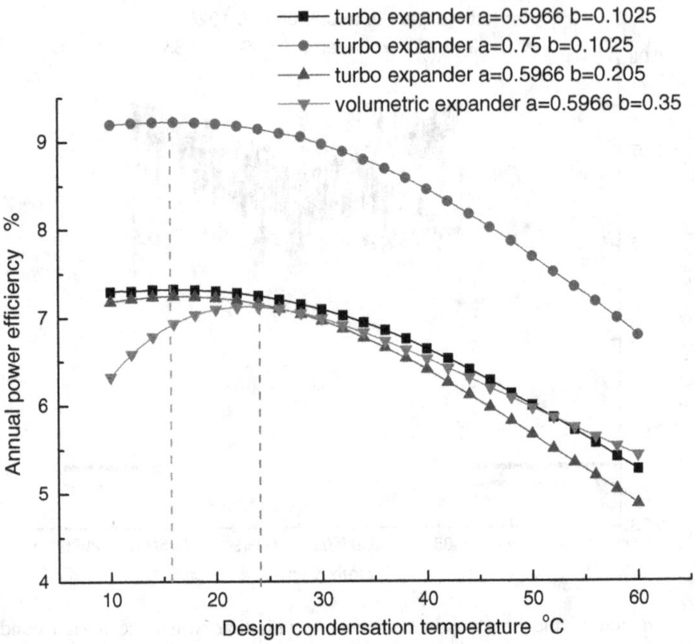

Fig. 4.18 Variation of the annual power efficiency with the design condensation temperature without CHP generation

7.32, 9.22, 7.24 and 7.13 %. Both the optimum design temperature and efficiency differ from those with CHP generation.

Figure 4.19 shows the variation of the optimum design and annual average operating condensation temperatures with the ratio of times of the SPG and CHP generation. A ratio of zero means the ORC works under the complete CHP mode through the year. The increment in the ratio means a longer time for the SPG mode starting on January 1st. The corresponding annual power efficiency is shown in Fig. 4.20. At a given ratio, the optimum design condensation temperature in the presence of the volumetric expander is higher than the annual average operating condensation temperature. But in the case of the turbo expander, it is lower. The maximum difference between the optimum design temperatures for the volumetric and turbo expanders occurs in neither the complete CHP mode nor the complete SPG mode.

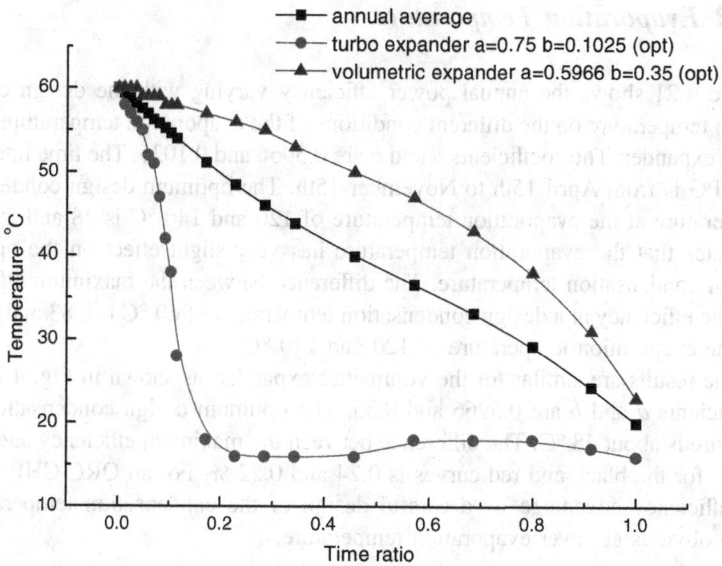

Fig. 4.19 Variations of the optimum design and annual average operating condensation temperatures with ratio of operation times of the SPG and CHP modes

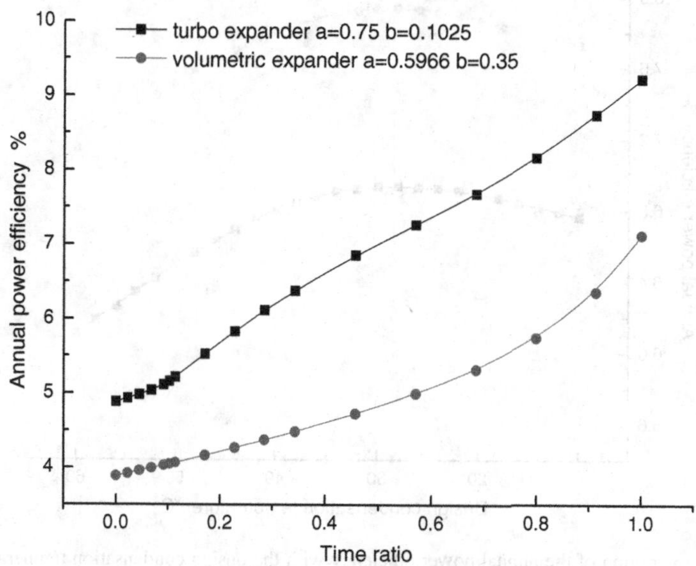

Fig. 4.20 Variation of the annual power efficiency with ratio of operation times of the SPG and CHP modes when the design condensation temperature is optimal

4.5.3 Evaporation Temperature

Figure 4.21 shows the annual power efficiency varying with the design condensation temperature on the different conditions of the evaporation temperature for the turbo expander. The coefficients a and b are 0.5966 and 0.1025. The time interval of the SPG is from April 15th to November 15th. The optimum design condensation temperature at the evaporation temperature of 120 and 140 °C is 28 and 29 °C. It indicates that the evaporation temperature has very slight effect on the optimum design condensation temperature. The difference between the maximum efficiency and the efficiency at a design condensation temperature of 60 °C is 0.83 and 0.78 % for the evaporation temperature of 120 and 140 °C.

The results are similar for the volumetric expander, as shown in Fig. 4.22. The coefficients a and b are 0.5966 and 0.35. The optimum design condensation temperature is about 48 °C. The difference between the maximum efficiency and that at 60 °C for the black and red curves is 0.24 and 0.22 %. For an ORC-CHP system, the efficiency advantage by a careful design of the condensation temperature is more obvious at lower evaporation temperature.

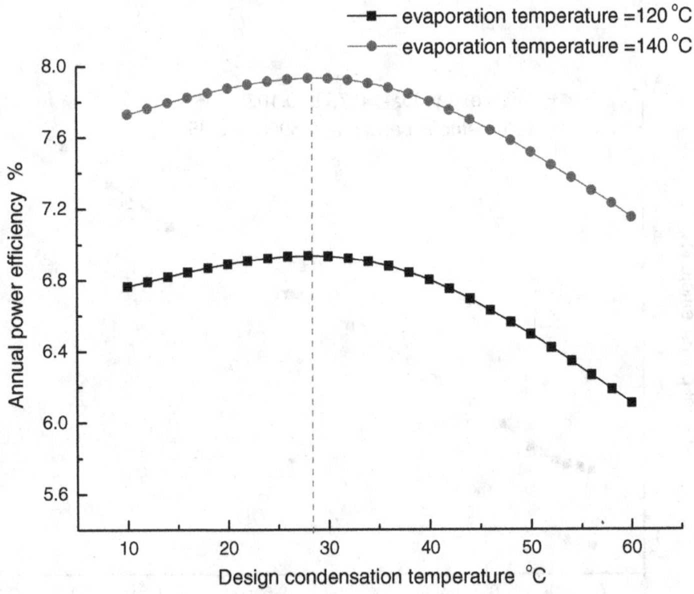

Fig. 4.21 Variation of the annual power efficiency with the design condensation temperature for the turbo expander at different evaporation temperatures

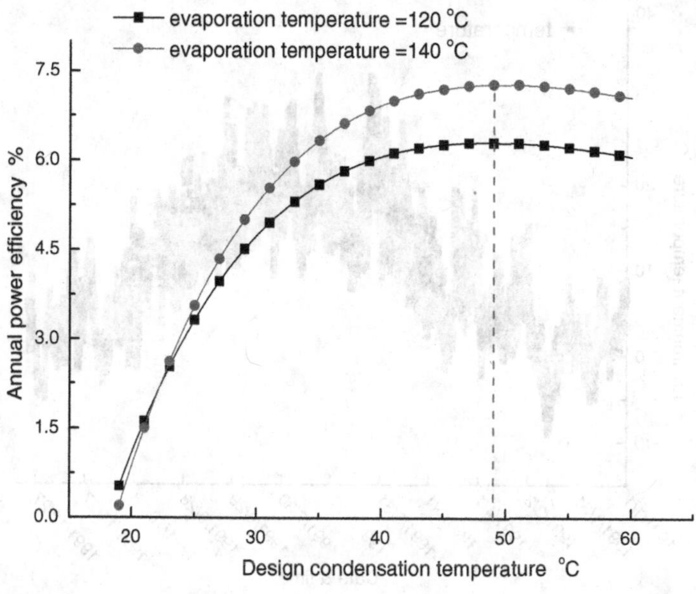

Fig. 4.22 Variation of the annual power efficiency with the design condensation temperature for the volumetric expander at different evaporation temperatures

4.5.4 Local Climate

The annual environment temperature differs from region to region. Figure 4.23 shows the case in Berlin. The monthly average environment temperature from January to December is 1.9, 0.3, 5.4, 8.3, 14.0, 17.6, 19.1, 18.5, 15.0, 10.2, 4.4, and 2.4. The annual average environment temperature is about 9.8 °C. Figure 4.24 displays the annual power efficiency varying with the design condensation temperature. The time interval of SPG mode is also from April 15th to November 15th. The optimum design condensation temperature for the turbo and volumetric expanders is about 20 and 45 °C. The annual average operating condensation temperature is about 35.7 °C. The shapes of these curves are similar with those for Hefei. However, with a lower annual environment temperature, the optimum design condensation temperature is lower and the corresponding annual average power efficiency is higher.

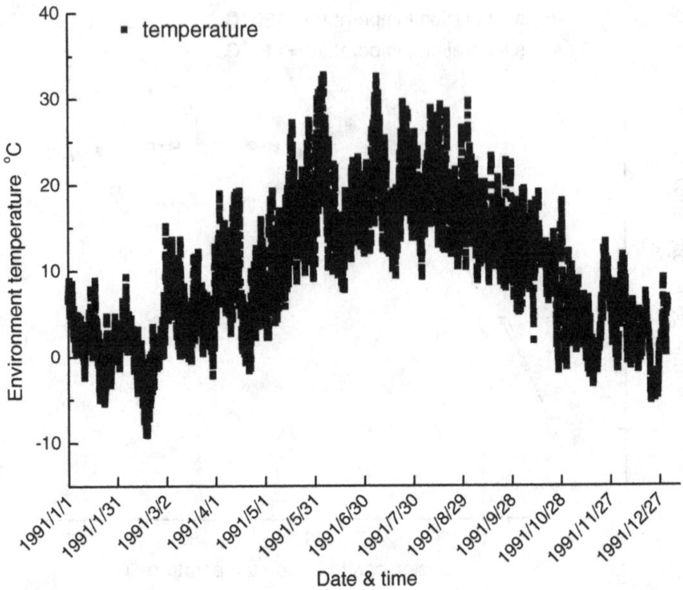

Fig. 4.23 Variation of the annual environment temperature of Berlin in 1991

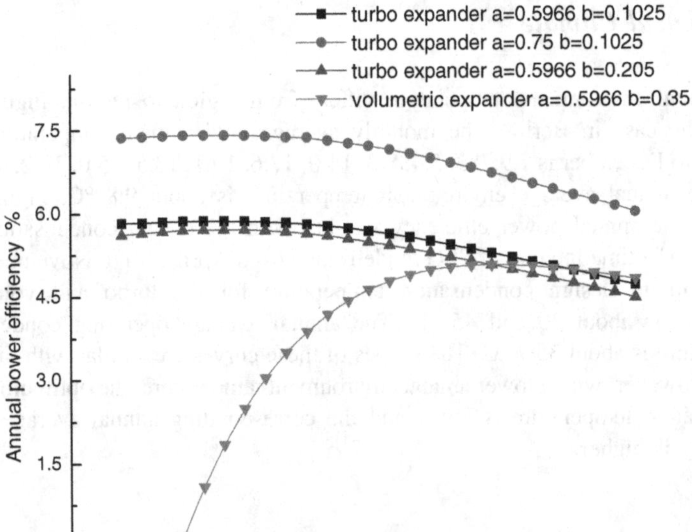

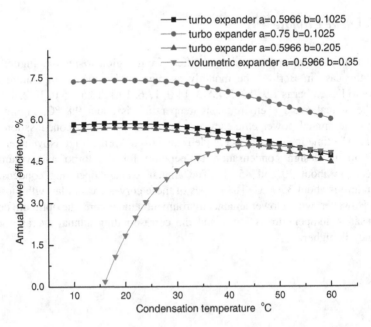

Fig. 4.24 Variation of the annual power efficiency with the design condensation temperature for Berlin

References

1. Bracco R, Clemente S, Micheli D, Reini M (2013) Experimental tests and modelization of a domestic-scale ORC (Organic Rankine Cycle). Energy 58:107–116
2. Colonna P, Bahamonde S (2013) Solar ORC turbogenerator for green-energy buildings. In: 2nd international seminar on ORC power systems, Rotterdam
3. Qiu GQ, Shao YJ, Li JX, Liu H, Riffat SB (2012) Experimental investigation of a biomass-fired ORC-based micro-CHP for domestic applications. Fuel 96:374–382
4. Twomey B, Jacobs PA, Gurgenci H (2013) Dynamic performance estimation of small-scale solar cogeneration with an organic Rankine cycle using a scroll expander. Appl Therm Eng 51:1307–1316
5. Lecompte S, Huisseune H, van den Broek M, De Schampheleire S, Paepe De M (2013) Part load based thermo-economic optimization of the Organic Rankine Cycle (ORC) applied to a combined heat and power (CHP) system. Appl Energy 111:871–881
6. Noussan M, Abdin CG, Poggio A, Roberto R (2013) Field operational analysis of an existing small size biomass-fired ORC unit. In: 2nd international seminar on ORC power systems, Rotterdam
7. Erhart TG, Eicker U, Infield D (2013) Influence of condenser conditions on organic Rankine cycle load characteristics. J Eng Gas Turbines Power 135:042301–042309
8. Cho SY, Cho CH, Ahn KY, Lee YD (2014) A study of the optimal operating conditions in the organic Rankine cycle using a turbo-expander for fluctuations of the available thermal energy. Energy 64:900–911
9. Zhang J, Zhou Y, Wang R, Xu J, Fang F (2014) Modeling and constrained multivariable predictive control for ORC (Organic Rankine Cycle) based waste heat energy conversion systems. Energy 66:128–138
10. Ibarra M, Rovira A, Alarcón-Padilla D, Blanco J (2014) Performance of a 5 kWe organic Rankine cycle at part-load operation. Appl Energy 120:147–158
11. Wang JF, Yan ZQ, Zhao P, Dai YP (2014) Off-design performance analysis of a solar-powered organic Rankine cycle. Energy Convers Manag 80:150–157
12. Manente G, Toffolo A, Lazzaretto A, Paci M (2013) An organic Rankine cycle off-design model for the search of the optimal control strategy. Energy 58:97–106
13. Seok Hun Kang (2012) Design and experimental study of ORC (organic Rankine cycle) and radial turbine using R245fa working fluid. Energy 41:514–524
14. Walnum HT, Rohde D, Ladam Y (2012) Off-design analysis of ORC and CO_2 power production cycles for low-temperature surplus heat recovery. Int J Low Carbon Technol 0:1–8
15. Walsh PP, Fletcher P (2004) Gas turbine performance, 2nd edn. Wiley, New York
16. Dixon SL, Hall CA (2010) Fluid mechanics and thermodynamics of turbomachinery, 6th edn. Elsevier Inc., Butterworth-Heinemann
17. Pump data booklet (2012) http://www.grundfos.dk/web/homevn.nsf/GrafikOpslag/IO_NB.NBG/File/NB,20NBG.pdf
18. Avadhanula VK, Lin CS (2014) Empirical models for a screw expander based on experimental data from organic Rankine cycle system testing. J Eng Gas Turbines Power 136:062601–062608
19. Lemort V, Quoilin S, Cuevas C, Lebrun J (2009) Testing and modeling a scroll expander integrated into an organic Rankine cycle. Appl Therm Eng 29:3094–3102
20. Zhu Y, Jiang L, Jin V, Lijun Y (2014) Impact of built-in and actual expansion ratio difference of expander on ORC system performance. Appl Therm Eng 71:548–558
21. Vlasic EP, Girgis S, Moustapha SH (1996) The design and performance of a high work research turbine. ASME J Turbomach 118:792–799
22. Woinowsky KM, Lavoie JP, Vlasic EP, Moustapha SH (1999) Off-design performance of a single-stage transonic turbine. J Turbomach 121:177–183
23. Li J, Pei G, Li Y, Wang D, Ji J (2012) Energetic and exergetic investigation of an organic Rankine cycle at different heat source temperatures. Energy 38:85–95

24. Pei G, Li J, Li Y, Wang D, Ji J (2011) Construction and dynamic test of a small scale organic Rankine cycle. Energy 36:3215–3223
25. Li J, Pei G, Ji J, Bai X, Li P, Xia L (2014) Design of the ORC condensation temperature with respect to the expander characteristics for domestic CHP applications. Energy 77:579–590
26. Glassman JA (1993) Estimating turbine limit load. NASA contractor report 191105, April 1993. http://ntrs.nasa.gov/archive/nasa/casi.ntrs.nasa.gov/19930016694_1993016694.pdf
27. http://apps1.eere.energy.gov/buildings/energyplus/weatherdata_about.cfm?CFID=734298&CFTOKEN=e51cf336ef4e3594-E6C19F62-A5E1-4E6F-480668E0A7E33ACD&jsessionid=5583799D8812EF0D3B5CC4D74C1B5070.eere,2011-2-2

Chapter 5
Conclusion and Future Work

This work outlines the perspective on solar ORC technology. The ORC-based low-medium solar thermal power generation has advantages of easer realization of heat storage, avoidance of complicated tracker, ability to scale down for decentralization applications, good cooperation with biomass energy, etc.

Structural optimization is important to a stable, flexible, efficient, and cost-effective solar ORC. Three types of systems are proposed: solar ORC with CDVG, solar ORC with PV module and osmosis-driven solar ORC. The solar ORC with CDVG using two-stage collectors and PCMs has many advantages as higher heat collection efficiency, better reaction to change in solar radiation and lower cost etc., when compared with the conventional system. The solar ORC with PV module seems feasible in regard to the low temperature coefficient and low cost of amorphous silicon solar cells, and good performance of ORC in low temperature application. It can produce more electricity per unit surface area than side by side PV panels and CPC collectors with ORC. The osmosis-driven solar ORC using the semi-permeable membrane seems suitable for small scale tri-generation. The alternative working fluid and membrane have strong technical background in absorption refrigeration, water desalination and purification, which strengthens the feasibility of the system.

A detailed study of the ORC performance under variable condensation temperature is conducted. The minimum condensation temperature is restricted by the environment temperature of about 20 °C and the maximum is about 50 °C for the CHP application. Compared with the previous works dealing with the static condensation temperature, this work provides a novel demonstration of the feasibility of small scale ORC-based CHP generation with low grade heat. According to the test results, the expander benefits from the low pressure ratio in the CHP generation. No appreciable increment in the expander losses has been observed at low pressure ratio. And the expander efficiency increases with the decrement in the pressure ratio, reaching the maximum value of about 0.54 at a pressure ratio of 2.6. Regarding a pressure ratio as low as 2.6 simple expander of only one-stage expansion is competent for the ORC-based CHP generation.

J. Li, *Structural Optimization and Experimental Investigation of the Organic Rankine Cycle for Solar Thermal Power Generation*, Springer Theses,
DOI 10.1007/978-3-662-45623-1_5

Aside from the expander, the ORC system reaches a higher degree of thermodynamic perfection when operating in the CHP mode. With a hot side temperature of about 100 °C and a cold side temperature of about 50 °C, the exergetic efficiency of the ORC is 57.8 % and the power efficiency is 3.5 %. The small scale ORC-based CHP generation seems feasible from the viewpoints of both cost-effectiveness and thermodynamic effectiveness.

The detailed investigation shows that with the increment in the condensation temperature, the ratio of the exergy destruction in the evaporator and condenser to the total system destruction becomes smaller while the ratio for the expander and pump becomes higher. The results indicate as the condensation temperature increases the thermodynamic process of R123 in the heat exchangers gets closer to isothermal heating/cooling, and the expander and pump play more important role in the cycle for heat to power conversion.

The non-constant efficiency of ORC components especially the expander in variable operation necessitates a careful design of the condensation temperature. According to the current and previous experiment results, the turbo and volumetric expanders behave distinctly when the pressure ratio deviates from the design. For the former, the efficiency variation is slight when the pressure ratio is lower than the design, making the left part of the efficiency curve much smoother than the right part. For the latter, the efficiency drops sharply at low pressure ratio.

A close view of the variation of the ORC power efficiency with the condensation temperature is presented. It is shown that the ORC power efficiency first increases and then decreases when the condensation temperature drops from the design value. Owing to the trade-off between the temperature difference driving the ORC and the expander efficiency, there is a threshold condensation temperature at which the ORC efficiency reaches the maximum. The threshold condensation temperature is important to the operation strategy of the ORC. It can offer significantly higher ORC efficiency than the design. In the meantime, measures shall be taken to prevent the condensation temperature declining below the threshold.

The optimum design condensation temperature is determined comprehensively by the expander characteristics, operating mode, evaporation temperature, and annual weather conditions. Among these factors, the expander characteristics and the duration of no-heating operation in a year are most crucial issues. On the given conditions, the optimum design condensation temperature for the turbo expander differs from that for the volumetric expander. Generally, the optimum design value is lower than the annual average operating value for the turbo expander, while it is higher than the average operating value for the volumetric expander.

In summary, the proposed solar ORCs in this work have advantages. The experiment results and design methodologies of the ORC are useful for solar CHP applications. The future work will be focused on building a demonstrating solar ORC system of tens of kW. The medium temperature solar collector and the small expander will play the key role in the development of such a system. The PTCs and CPCs of low concentration ratio are capable to convert solar irradiation to medium-grade heat with adequate efficiencies. The PTCs have been mass-produced by many manufactures, but require a tracking system. The CPCs are expected to reach a higher degree of maturity in the coming years. Turbo-expanders and volumetric expanders are both competitive in the small-scale ORC power generation. However, the turbo-expanders have to overcome the hurdle of high rotation speed at low power. There will be several solutions including the radial outflow and oil-free magnetic bearings technologies. The radial outflow configuration is characterized by multiple stages, radial outflow with or without last axial stage, overhung, oil sealed, and rolling bearings. Due to the single disk arrangement, it can accompany a high number of stages without rotor-dynamic limitation and offers low rotation speed. The magnetic bearings replace conventional oil-lubricated bearings, eliminating high friction losses and mechanical wear. By connecting the expander with the generator directly, the gearbox becomes unnecessary. All these contribute to superior power conversion and delivery efficiencies. Volumetric expanders, on the other hand, seem advantageous in low power application but need to reach a higher commercialized level.

Printed in the United States
By Bookmasters